COMMERCIAL ENERGY AUDITING
REFERENCE HANDBOOK

COMMERCIAL ENERGY AUDITING
Reference Handbook

Steve Doty, PE, CEM

THE FAIRMONT PRESS, INC.

CRC Press
Taylor & Francis Group

Library of Congress Cataloging-in-Publication Data

Doty, Steve.
    Commercial energy auditing reference handbook / Steve Doty.
        p. cm.
    Includes index.
    ISBN 0-88173-567-1 (alk. paper) -- ISBN 0-88173-568-X (electronic book) -- ISBN 1-4200-6111-9 (distributor)
        1. Buildings--Energy conservation. 2. Energy auditing. I. Title.

TJ163.5.B84D68 200
658.2'6--dc22'

2007035073

Commercial Energy Auditing Reference Handbook / Steve Doty.
©2008 by The Fairmont Press. All rights reserved. No part of this publication may be reproduced or transmitted in any form or by any means, electronic or mechanical, including photocopy, recording, or any information storage and retrieval system, without permission in writing from the publisher.

Published by The Fairmont Press, Inc.
700 Indian Trail
Lilburn, GA 30047
tel: 770-925-9388; fax: 770-381-9865
http://www.fairmontpress.com

Distributed by Taylor & Francis Ltd.
6000 Broken Sound Parkway NW, Suite 300
Boca Raton, FL 33487, USA
E-mail: orders@crcpress.com

Distributed by Taylor & Francis Ltd.
23-25 Blades Court
Deodar Road
London SW15 2NU, UK
E-mail: uk.tandf@thomsonpublishingservices.co.uk

Printed in the United States of America
10 9 8 7 6 5 4 3 2 1

0-88173-567-1 (The Fairmont Press, Inc.)
1-4200-6111-9 (Taylor & Francis Ltd.)

While every effort is made to provide dependable information, the publisher, authors, and editors cannot be held responsible for any errors or omissions.

Printed on recycled paper.

Table of Contents

Introduction xi
To the Reader xiii
User Guide xv
Suggested References xvii

SECTION I—SPECIFIC INFORMATION 1
Chapter 1 - Benchmarking 3
    Differentiating by Energy Source 3
    Energy Use Intensity 4
    Limitations of EUI 4
    Calculating EUI 4
    Mixed EUI 5
    Production EUI 5
    Energy End Use Distribution 6
    Energy Cost as a Percent of Total Operating Cost 6
    Limitations of Using Benchmark Data 6

Chapter 2—Analyzing Energy Use Graphs 9
    Intuitive Information 9
    Comparative Information 10
    Weather Dependence 13
    Load Factor Effect on Energy Cost 14
    Business Volume (Production Rates) 16
    Savings Opportunities 19

Chapter 3—Energy Saving Opportunities by Business Type 21
    Apartment Buildings/Multi-Family/Dormitories 21
    Churches/Worship 24
    Data Centers 26
    Education—Colleges and Universities 28
    Education—Schools K-12 31
    Food Sales—Grocery Stores 34
    Health Care—Hospital 36
    Health Care—Non Hospital 38
    Laundries—Commercial 41
    Libraries/Museums 42
    Lodging/Hotels/Motels 44
    Office Buildings 47
    Food Service 50

Retail/Sales	53
Warehouses	56
Boilers	58
Chillers	58
High Rise	59
Pools	59
Ice Rinks	61

Chapter 4—Manufacturing ... 63
- Process Analysis ... 63
- Example Manufacturing Process Flow Chart ... 65
- Desire for Energy Use to Follow Production Rates ... 65
- Primary Energy Use Sources ... 65
- Production Scheduling ... 66
- Maintenance ... 66
- Controls ... 66
- Some Common ECMs for Manufacturing ... 67

Chapter 5—ECM Descriptions ... 73
- ECM Descriptions—Envelope ... 73
- ECM Descriptions—Lighting ... 75
- ECM Descriptions—HVAC ... 77
- ECM Descriptions—Swimming Pools ... 93
- ECM Descriptions—Heat Recovery ... 95
- ECM Descriptions—Thermal Storage (TES) ... 99
- ECM Descriptions—Electrical ... 102
- ECM Descriptions—Other ... 103

Chapter 6—Utility Rate Components ... 105
- Electric ... 105
- Gas ... 109

Chapter 7—Automatic Control Strategies ... 111
- Cost/Benefit Ratio for Control System ECMs ... 111
- Control System Application Notes ... 113
- Lighting Control Strategies—Basic ... 113
- HVAC Control Strategies—Basic ... 114
- Lighting Control Strategies—Advanced ... 127
- HVAC Control Strategies—Advanced ... 127
- Other Control Sequences ... 134
- Other Ways to Leverage DDC Controls ... 135

Chapter 8—Building Operations and Maintenance ... 137

Facility Repair costs 137
Maintenance Value 137
Poor Indoor Comfort and Indoor Air Quality Costs 137
Productivity Value 138
Maintenance Energy Benefits 139

Chapter 9—Quantifying Savings 153
    Establishing the HVAC Load Profile 154
    Load-Following Air and Water Flows vs. Constant Flow (VFD Benefit)...159
    VAV System Fan Savings Reduction for Maintaining
        Downstream Pressure 165
    Supply Air Reset vs. Reheat – Constant Volume 166
    Supply Air Reset with VAV vs. Increased Fan Energy 167
    Condenser Water Reset vs. Constant Temperature 176
    Chilled Water Reset for Variable Pumping vs. Increased Pump Energy ...181
    Water-Side Economizer vs. Chiller Cooling 184
    Higher Efficiency Lighting vs. Existing Lighting 193
    Higher Efficiency Motors vs. Existing Motors 196
    Higher Efficiency Chiller vs. Existing Chiller 200
    Higher Efficiency Boiler vs. Existing Boiler 201
    Hot Water Reset From Outside Air vs. Constant Temperature 204
    Reduce Air System Friction Losses – Constant Volume 208
    Computer Modeling 210
    Measurement and Verification (M&V) 212

Chapter 10—Sustaining Savings 215
    Tendency for initial savings to deteriorate 215
    Maintaining initial savings 215
    Checklist for service access and operations 216

SECTION II—GENERAL INFORMATION 219
Chapter 11—Mechanical Systems 221
    Relative Efficiency of Air Conditioning Systems 221
    Glossary of Basic HVAC System Types 221
    Water-Cooled vs. Air Cooled 224
    Thermal Energy Transport Notes 224
    Chillers 225
    Part Load Chilled Water System Performance 228
    Cooling Towers and Evaporative Fluid Coolers 232
    Dry Coolers 234
    Electronic Expansion Valves 234
    Air and Water System Resistance 235

Fan/Pump Motor Work Equation 236
Fan and Pump Efficiencies, and Belt Drive Efficiencies 237
Fan and Pump Throttling Methods 238
Thermal Break-Even Concept for Buildings...239
Air-Side Economizer 240
Computer Room Air Conditioning (Data Center)......................................244
Cooling Energy Balance for Heat Producing Equipment..........................246
Humidifiers 246
Kitchen Grease Hoods (Type 2) 249
Heat Pumps..250
Refrigeration Cycle..253
Evaporative Cooling 255
Spot Cooling 262
VAV Reheat Penalty 263
Glycol vs. Efficiency 263
Cost of Ventilation 267
Simultaneous Heating and Cooling..268

Chapter 12—Motors and Electrical Information 269
Motor Efficiencies..269
Voltage Imbalance 270
Sources of Electric Motor Losses 273
Power Factor Correction Capacity Quick Reference Chart......................274

Chapter 13—Combustion Equipment and Systems 277
Cost of Steam 277
Range of Combustion Efficiency 277
Combustion Efficiency Nomograph 278
Boiler Heating Output When Only
    Heating Surface Area is Known 279
Boiler Standby Heat Loss (Boiler Skin Loss) at Full Load 279
Estimated Losses from Boiler Short Cycling 280
Dampers and Boiler Isolation Valves. 281
Estimated Savings of Steam System Improvement 284

Chapter 14—Compressed Air 289
Estimated Savings of Compressed Air Improvements 289

Chapter 15—Fan and Pump Drives 293
Fan Drive Efficiency Comparison Diagram 293
Cog Belts Instead of Standard V-Belts 294
Variable Speed Drive Considerations..294

Chapter 16—Lighting 297
    Lighting terms 298
    Dimming 298
    Light Colored Surfaces 299
    General Lighting Information 300
    Typical Recommended Lighting Levels 302
    Occupancy Sensor Energy Savings 304

Chapter 17—Envelope Information 305
    BLC Heat Loss Method 305
    R-Value Reduction from Stud Walls 306
    Glazing Properties 307
    Infiltration 310

Chapter 18—Water and Domestic Water Heating 311
    Water Consumption for Water-Cooled Equipment 311
    Blowdown vs. Cycles of Concentration 311
    Plumbing Water Points of Use 312
    Domestic Water Heaters 312
    Domestic Water Heater Stand By Losses 313

Chapter 19—Weather Data 315
    Degree Days 315
    Bin Weather Data 316
    Weather by Days and Times 317

Chapter 20—Pollution and Greenhouse Gases 321
    Emission Conversion Factors by Region 321
    Greenhouse Gas Relationship to Energy Use 321
    Emissions from Burning Fossil Fuels. 324
    Pollution-conversion to Equivalent Number of Automobiles 325
    Other Environmental Considerations 326

Chapter 21—Formulas and Conversions 327
    Common Energy Equations 327
    Other Useful Formulas 337
    Energy Conversion Factors 339

APPENDIX 341
Glossary of Terms 341
Conflicting ECMs and 'Watch Outs' 345
Types of Energy Audits 348

Pressure-Temperature Charts for Refrigerants 349
Cost Estimating—Accuracy Levels Defined 354
Simple Payback vs. Internal Rate of Return (IRR) 355
Heat Loss from Uninsulated Hot Piping and Surfaces 358
Heat Loss from Insulated Piping 358
Duct Fitting Loss Coefficients 359
Evaporation Loss from Water in Heated Tanks 361
Bin Weather Data for 5 Cities (dry bulb) 362
Altitude Correction Factors at Different Temperatures 363
Energy Use Intensity (EUI)—per SF—by Function and Size 364
Energy Use Intensity (EUI)—per SF—by Function and Climate Zone 365
Energy Use Intensity (EUI)—per SF—by Function, Climate Zone and Size 365
Energy Use Intensity (EUI)—per SF—Measured at One Data Center 370
CBECS Climate Zone Map 370
Building Use Categories Defined (CBECS) 371
Operating Expenses: Percent that are from Utility Costs 375
Energy Use Intensity (EUI)—per SF—for Some Manufacturing Operations 377
Energy Use Intensity (EUI)—in Production Units—Some Mfg. Operations 378
ASHRAE 90.1—Items Required for All Compliance Methods 379
Top 15 Emerging Technologies—2002 (DOE) 384
HVAC Retrofits for the Three Worst Systems 385
Service Life of Various System Components 388
Equating Energy Savings to Profit Increase 390
Integrated Design Examples 394
Energy Audit Approach for Commercial Buildings 395
Energy Audit Look-for Items 397
Sample Pre-audit Customer Questionnaire 401
Facility Guide Specification: Suggestions to Build in Energy Efficiency 418
ASHRAE Psychrometric Charts 1-5 437

INDEX 443

Introduction

This handbook is a valuable collection of reference material related to commercial energy audit-ing, written for practicing energy professionals. The main goal of this book is to **increase audit effectiveness so that more audits become projects.** It includes proven solutions and practical guidelines, as well as the essential information needed for winning energy audits and projects.

The motivation for writing this book:

- Having accurate facts and figures in **one convenient location** will accelerate the research phase of each audit while maintaining quality.

- The benefit of **hands-on practical experience** is found throughout this book, including things that work and things to watch out for.

- Audit strategies that are **tailored to each unique business sector** are more effective than general guidelines, and the first step is to learn where the energy is used in the business. The analogy is saving gasoline at your home: begin with the automobile, not the lawn mower.

- **Automatic control systems** are often underutilized and represent untapped potential for savings. To leverage this technology, a separate section provides specific automatic control applications.

- **Quantifying savings** is a skill that is essential for success, but one that is treated lightly in many texts and so a section is devoted to identifying the fundamentals at work and how to quantify them.

- Identifying **maintenance** activities that reduce energy use represent immediate low cost savings opportunities for customers.

- The first step to reducing energy use in the commercial building population is to encourage **changes in how *new* buildings are built** and operated, and there is a section on this subject. By curbing the energy use of new buildings, we can eventually catch up.

Steve Doty

To the Reader

Here are the two main ways I use to find information in this book, with examples.

Method 1: (Table of Contents or TOC): Identify the general category to narrow the search, then browse the sub categories using the table of contents and appendix.

Method 2 (Index): Identify unique key words or phrases for your subject matter. Using the index, find the key word directly.

Example: What are typical measures that make sense to a school?
(TOC): Look in the section "Specific Information" and find "Energy Saving Opportunities by Business Type"; then find "Education—Colleges and Universities" or "Education—Schools K-12"
(Index): Look for "schools" or "universities" or "K-12."

Example: What is the conversion between ton-hours and kWh?
(TOC): Look in the section "General Information" and find "Formulas and Conversions"; then find "HVAC Formulas and Conversions."
(Index): Look for "ton-hours."

Example: What is a reasonable energy use per SF for an office building?
(TOC): Look in the "Appendix" and find "Energy Use Intensity (EUI) —Per SF—By Function and Size."
(Index): Look for "EUI" or "energy use intensity."

Another resource within the book is a glossary of terms in the Appendix.

User Guide

This handbook contains a great deal of information in condensed form and its success will depend in large part on the ability of the reader to navigate through the material. A condensed table of contents below shows the basic organization of subject matter. In the full table of contents, each item is expanded as shown in this example. There is also a generous index of key words for quick access.

SECTION I. SPECIFIC INFORMATION
Chapter 1   Benchmarking
Chapter 2   Analyzing Energy Use Graphs
Chapter 3   *Energy Saving Opportunities by Business Type*
Chapter 4   Manufacturing
Chapter 5   ECM Descriptions
Chapter 6   Utility Rate Components
Chapter 7   Automatic Control Strategies
Chapter 8   Building Operations and Maintenance
Chapter 9   Quantifying Savings
Chapter 10 Sustaining Savings

SECTION II. GENERAL INFORMATION
Chapter 11 Mechanical Systems
Chapter 12 Motors and Electric Information
Chapter 13 Combustion Equipment and Systems
Chapter 14 Compressed Air
Chapter 15 Fan and Pump Drives
Chapter 16 Lighting
Chapter 17 Envelope Information
Chapter 18 Water and Domestic Water Heating
Chapter 19 Weather Data
Chapter 20 Pollution and Greenhouse Gases
Chapter 21 Formulas and Conversions

```
Apartment Buildings/Multi-Family/
   Dormitories
Churches/Worship
Data Centers
Education—Colleges and Universities
Education—Schools K-12
Food Sales—Grocery Stores
Health Care—Hospital
Health Care—Non Hospital
Laundries—Commercial
Libraries/Museums
Lodging/Hotels/Motels
Office Buildings
Food Service
Retail/Sales
Warehouses
Boilers
Chillers
High Rise
Pools
Ice Rinks
```

APPENDIX

INDEX

Suggested References

This is a condensed collection of information for professionals engaged in energy engineering. It is mostly without explanatory material and relies on readers with an understanding of the underlying principles and applications. It will also serve as a companion reference for other texts on individual subjects. The following are suggested sources of additional information and detail.

Energy Engineering and Energy Management
Energy Management Handbook, Turner/Doty, Fairmont Press
Handbook of Energy Engineering, Thumann/Mehta, Fairmont Press
Information Technology for Energy Managers, Capehart, Fairmont Press

HVAC
Principles of Heating, Ventilating and Air-Conditioning, ASHRAE
Pocket Guide for Air-Conditioning, Heating, Ventilation and Refrigeration, ASHRAE

Lighting
Lighting Management Handbook, DiLouie, AEE
IES Lighting Ready Reference, IES

Automatic Controls
Fundamentals of HVAC Control Systems, Taylor, ASHRAE
Optimization of Unit Operations, Liptak, Chilton

Section I
Specific Information

Chapter 1

Benchmarking

GENERAL

A good place to begin an energy audit is to compare the use of the facility with similar facilities. For the purpose of this book, benchmarking refers this initial comparative step. It is a rough gage of whether the existing use is more than, less than, or about the same as what would be expected.

It provides a basic indicator of potential for savings. For **example**, if the annual energy use per SF is 25% higher than a trusted benchmark value, then a 25% savings for the facility may be a reasonable goal. Similarly, if the energy use is about equal to the available benchmark, a customer desiring to reduce energy use by 50% will likely require extreme measures and may not be practical.

The comparison may be in any meaningful and available units. Often customers or performance contractors are interested in dollars instead of energy units, since dollars are what drive the business. This is understandable, however benchmarking in dollars introduces additional variables of the cost of energy—and this can vary substantially by area and over time. For this reason, benchmarks that are in terms of energy units instead of dollars introduce less uncertainty and are easier to work with.

Examples:

Energy use per SF per year	(kBtu/SF-yr)
Energy use per part produced	(kBtu/part)
Energy use per ton of material processed	(kBtu/ton)
Energy use per month per hotel guest	(kBtu/month-guest)
Energy use per 100 meals served	(kBtu/100 meals)

DIFFERENTIATING BY ENERGY SOURCE

Benchmarking is possible for electric use vs. natural gas use or other fuels, however this introduces more variables. For **example**, if one building is heated with electric resistance, and one is heated with natural gas, the benchmarks would be different with electric and gas energy evaluated separately. But if the benchmark is simply Btus of energy, then buildings heated by any means would be measured against the same standard.

ENERGY USE INTENSITY

The term *energy use intensity* (EUI) is a common measure of energy consumption in commercial buildings. It is synonymous with energy use index and energy use intensiveness. It is derived from the total consumed energy per year, divided by the square foot area of the building. It represents how concentrated the energy use is within the building. Buildings with high EUI also tend to be self heating. A variety of EUI data is provided in the **Appendix**.

LIMITATIONS OF EUI

EUI is a simple measure, with limitations. For **example**, building volume is not considered. A warehouse with high stacked storage and high ceilings will have increased energy use through envelope heat loss/heat gain compared to shorter warehouse building with less wall area.

CALCULATING EUI

Note: for these examples, natural gas is used. The same method applies to other fuels.

Single EUI
For defined business sectors with available data, comparing actual to 'reasonable' energy intensiveness (Btu/SF-yr) is a logical first step.

Step 1: Convert one year's energy consumption data, gas and electric, to the common unit of kBtu (1000's of Btus).

    1 kBtu = 1000 Btu
    1 kWh = 3.413 kBtu

1 therm = 100,000 Btu
1 ccf= _____ Btu **
**This varies by locale, typically around 100,000 Btu/100 cubic feet (cf).
some as high as 1050/cf or 105,000 per ccf
some as low as 800 Btu/cf or 800,00 per ccf

Step 2: Determine building square footage that is occupied. Exclude garages, crawl spaces, breezeways, etc.

Step 3: Energy Use Intensiveness (EUI) = kBtu per year / SF

Step 4: Compare to available benchmark data.

See the **Appendix** for common EUI benchmark data.

MIXED EUI

It is common to create an "adjusted EUI" for facilities with defined constituent parts, using proportions.
Example: A commercial building is 70% lodging and 30% restaurant. From the **Appendix**, Lodging EUI=100; Restaurant EUI=231.
(0.7)*(100)+(0.3)*(231) = 139 kBtu/SF-yr combined (ans.)

PRODUCTION EUI

A similar approach can be used for manufacturing or other volume-related activity.

For example: If 500,000 screwdrivers are manufactured in a year, the energy cost per screwdriver can easily be calculated. In this example, energy is seen as an ingredient to the produced item, similar to steel or labor, and savings equate to reduced part cost and increased profit. Restaurants (meals per day) and hotels (average percent occupancy) are similar applications of production metrics.

OTHER BENCHMARKING METRICS
ENERGY END USE DISTRIBUTION

Sometimes called energy use "pies," these predict the percent of total use going to which function, such as heating, lighting, etc. This can be used as a benchmark if a constituent piece of energy use is known.

Example: If the domestic water heaters are electric and the only thing using natural gas is the heating system, then the natural gas bills represent the entire heating energy use. If this amount of energy represents 40% of total energy but the end use pie shows 20%, this may indicate excess heating energy is being used for some reason.

Chapter 3 shows energy end use distribution pie diagrams for some common business sectors.

ENERGY COST AS A PERCENT OF TOTAL OPERATING COST

For operating budgets, "utilities" is a line item and often most of that is energy. Each percent of savings on utilities translates to increased profit. This is shown in table form in the Appendix item: **"Equating Energy Savings to Profit Increase."** If the operating cost percentage is high compared to other similar companies, then profit will usually be proportionally lower.

Information on operating expenses is available from private industry groups, and also from the U.S. Census Bureau "Business Expenses" report. To get energy cost as a percent of total operating cost, divide the utility cost by the total operating expenses. Some typical energy cost percentages are shown in the **Appendix** item: **"Operating Expenses: Percent that are from Utility Costs"**

LIMITATIONS OF USING BENCHMARK DATA

While very useful, remember that benchmarks are averages. When a test value is significantly different than a benchmark, it prompts questions and may lead to opportunities. However when a test value is only slightly different than the benchmark, the conclusion may be only that the numbers are close. For **example**, if a building EUI is calculated at 102 kBtu/SF and the benchmark for comparison is 100 kBtu/SF, this is not proof positive that the building energy use is 2% higher than normal. This is also discussed in the **Appendix** item: "**Energy Use Intensity (EUI)—Per SF—By Function, Climate Zone, and Size.**"

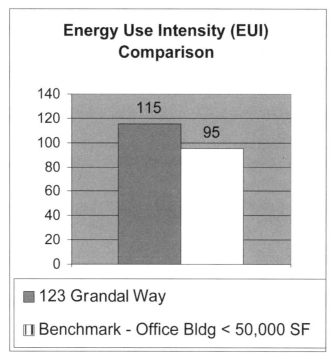

Figure 1-1. EUI Benchmark Comparison for an Office Building

Chapter 2

Analyzing Energy Use Graphs

GENERAL

Energy use profiles, by month, are very useful in most cases, especially when patterns or abnormalities suggest issues and possible energy saving opportunities. Studying these is a good initial step.

Graphing is also very effective for presenting complex information to customers.

Finally, graphing is an excellent tool for displaying the results of changes made—before and after. Graphing with two vertical axes can be insightful for things happening concurrently.

INTUITIVE INFORMATION

Some things are obvious from how they look.

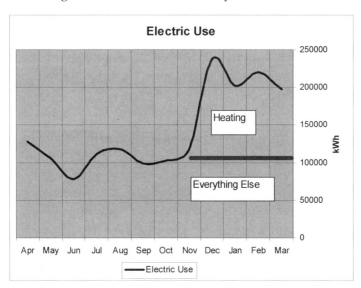

Figure 2-1a.

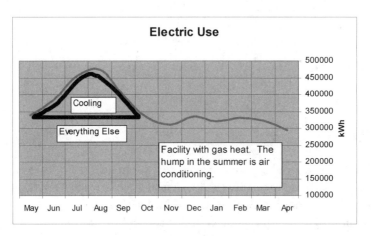

Figure 2-1b.

COMPARATIVE INFORMATION

Overlay the Same Periods of Time

Comparative information:
Showing consistent use year-to-year.

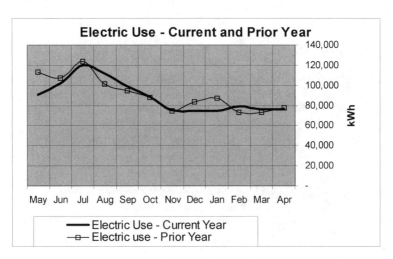

Figure 2-2a.

Analyzing Energy Use Graphs

Comparative information:
Showing energy usage creeping up year-to-year.

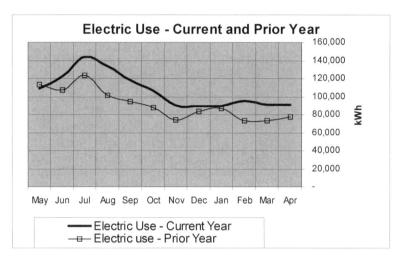

Figure 2-2b.

Comparative information:
Showing anomalies and prompting questions. This may have been a construction event or some other one-time event.

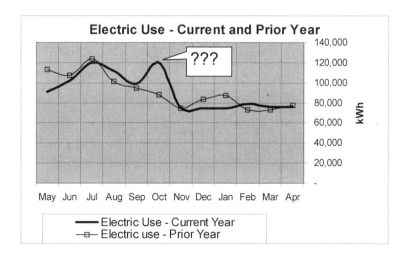

Figure 2-2c.

Comparative information:
Showing anomalies and offering explanations.

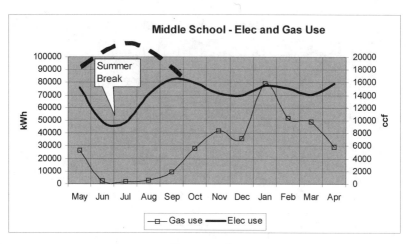

Figure 2-2d.

- Comparative information:
Showing results of projects.

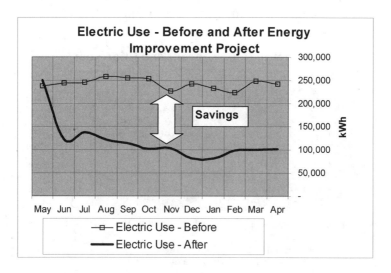

Figure 2-2e.

Comparative information:
Showing results of projects, baseline, expected, actual.

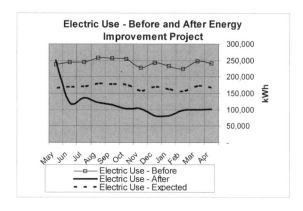

Figure 2-2f.

WEATHER DEPENDENCE

Weather Dependent Example

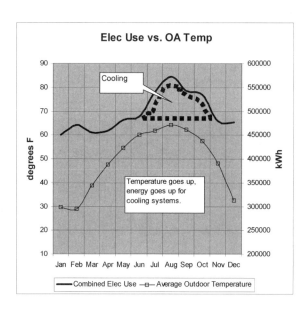

Figure 2-3a.

Weather Independent Example

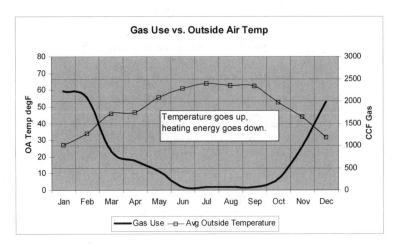

Figure 2-3b

LOAD FACTOR EFFECT ON ENERGY COST

Note the mirror image effect of poor load factor on utility cost. Low load factors (ratio of average to maximum demand) cause higher demand charges and increase the overall cost of power.

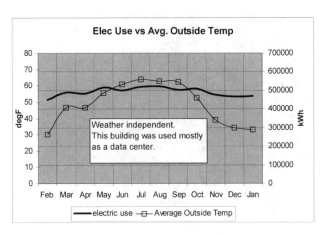

Figure 2-4a.

Changes that improve load factor, such as steady use throughout the day and avoiding brief periods of high electrical demand, can reduce electrical charges while energy use remains the same.

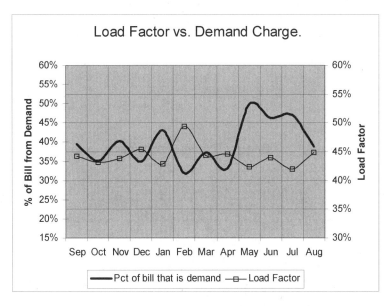

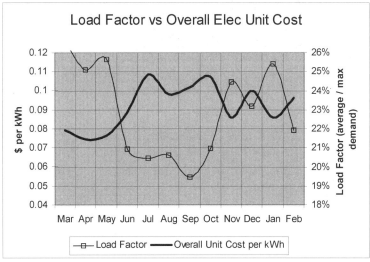

Figures 2-4b and 2-4c.

BUSINESS VOLUME (PRODUCTION RATES)

Data points are normalized, 100% being the most production and most energy during the graphed period. When energy use does not go down proportionally with production, energy use per unit of product output increases. This concept applies equally to manufacturing, hotels, restaurants, etc.

Example: Strong correlation (manufacturing)

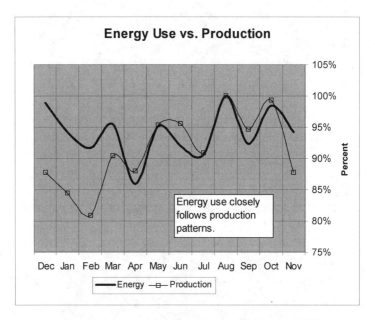

Figure 2-5a.

Analyzing Energy Use Graphs

Example: Weak correlation (manufacturing)

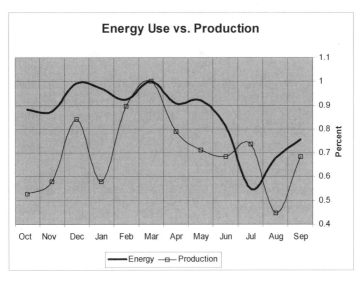

Figure 2-5b.

Hotel Example

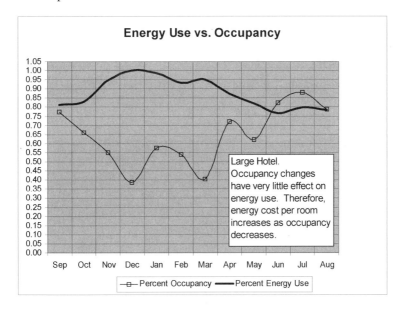

Figure 2-5c.

Energy cost per unit for this manufacturing facility increases with reduced output, from idling equipment energy use.

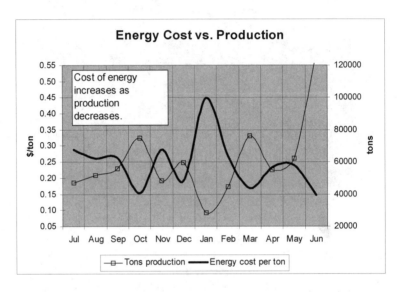

Figure 2-5d

Analyzing Energy Use Graphs

SAVINGS OPPORTUNITIES

Savings opportunity: Boiler running in summer, standby losses.

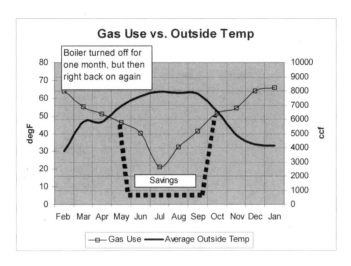

Figure 2-6a.

Savings opportunity: Overlapping heating and cooling.

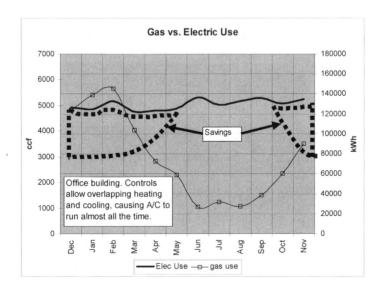

Figure 2-6b.

Savings opportunity: Overlapping heating and cooling.

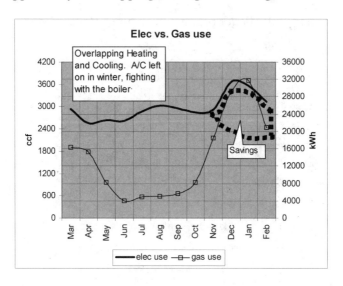

Figure 2-6c.

Savings opportunity: Economizer not functioning.

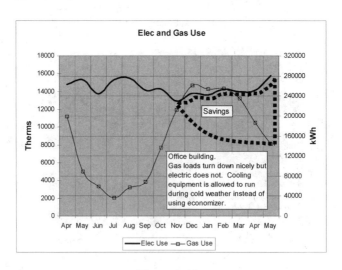

Figure 2-6d.

Chapter 3

Energy Saving Opportunities by Business Type

The dominant energy users are listed first, which are the drivers of utility costs and the most likely categories to bring the largest results. There will be other measures that can be suggested at each facility.

Common measures are organized in groups, which are:
- Controls
- Maintenance
- Low-cost/No Cost
- Retrofit—Or Upgrade at Normal Replacement
- May Only Be Viable During New Construction

APARTMENT BUILDINGS/MULTI-FAMILY/ DORMITORIES

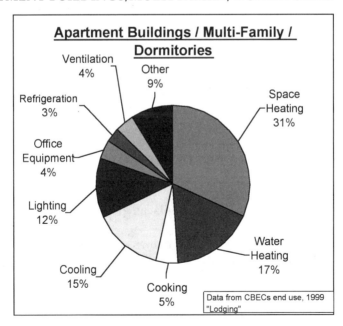

Figure 3-1. Energy End Use Distribution for Apartment Buildings/Mullti-family/Dormitories

Primary Energy Use Sources
- Heating and cooling
- Water heating
- Lighting

Controls
- Controls to lock out cooling below 50 deg F.
- Controls to lock out heating above 65 deg F.
- Reset boiler (HWS) based on outside air temp.
- Lower domestic hot water temperature to 120 deg F.
- Turn off domestic hot water re-circulation pumps at night.
- Occupancy sensors in amenity areas (fitness room, etc.).

Maintenance
- Calibrate controls every two years.
- Annual maintenance on all heat transfer surfaces, including good quality filters, cleaning coils, cleaning tubes, cleaning apartment refrigerator coils.
- Look for open windows in cold weather as signs of defective controls.
- Repair or replace defective zone control valves.

Low Cost/No Cost
- Thermostats with a minimum of 5 degrees deadband between heating and cooling.
- Low flow faucets and shower heads.
- Insulate bare hot water piping.
- Flyers to encourage energy conservation. Possible shared savings or picnic, etc.

Retrofit—Or Upgrade at Normal Replacement
Heating and Cooling
- Higher efficiency heating and cooling equipment.
- Increase roof insulation thickness to current energy code level as part of roof replacement.
- Window replacement.

Water Heating
- Condensing domestic water heater.
- Stack damper for gas-fired domestic water heater.

Lighting
- Higher efficiency lighting in common areas.
- Higher efficiency lighting in room areas, esp. to replace incandescent lighting.

Other
- Low water volume/high speed spin wash machines (E-Star).
- Energy Star Appliances on replacement.

May Only Be Viable During New Construction
- Envelope construction to match local residential energy code for insulation.
- Overhangs or other exterior shading at windows.
- Toilet exhaust on demand (light switch) instead of continuous.
- Low volume bathtubs.
- High performance glass on all windows.

Pool: See "Pools" this chapter.
If a boiler is used: See "Boilers" this chapter.
If a chiller is used: See "Chillers" this chapter.
If the building is over four stories in height: See "High Rise" this chapter.

CHURCHES/WORSHIP

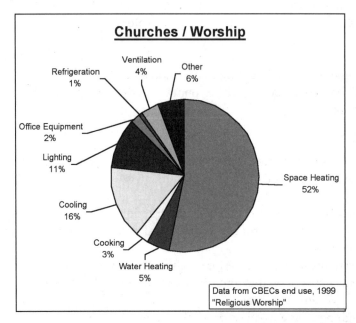

Figure 3-2. Energy End Use Distribution for Churches / Worship

Primary Energy Use Sources
- Heating and cooling
- Lighting

Controls
- Scheduled start-stop of HVAC equipment and lighting.
- Space temperatures set for 75 cooling/70 heating during services, and allowed to float higher and lower in other times, and at night.
- Morning warm-up or cool-down before services with outside air closed.
- Demand controlled ventilation for variable occupancy.
- For all roof openings to hoods and equipment not active in winter, dampers should be tightly closed during heating operation and when roof equipment is off.

Maintenance
- Calibrate controls every two years.

- Annual maintenance on all heat transfer surfaces, including good quality filters, cleaning coils, cleaning tubes.

Low Cost/No Cost
- Programmable thermostat.

Retrofit—Or Upgrade At Normal Replacement
Heating and Cooling
- Higher efficiency heating and cooling equipment.
- Anti-stratification fans for high bay, heated areas to move warm air to the floor.
- Reduce outside air to proper quantities if excessive.
- Increase roof insulation thickness to current energy code level as part of roof replacement.

Lighting
- Higher efficiency lighting
- Replace incandescent lighting with more efficient technology.

May Only Be Viable During New Construction:
- Passive shading elements for large glass areas.
- High performance low E glazing
- Exhaust heat recovery for high ventilation times.

If a boiler is used: See "Boilers" this chapter.

DATA CENTERS

Primary Energy Use Sources
- Computer equipment
- Cooling
- Lighting (distant third)

See Chapter 11—Mechanical Systems, "Computer Room Air Conditioning," "Proportions of CRAC Unit Energy to Total Computer

Table 3-1. Computer Data Center Energy Savings Opportunities
Source: "Energy Efficiency in Computer Data Centers," Doty, *Energy Engineering Journal*, Vol. 103, No. 5, 2006.

System Type	Measure	New or Retrofit	Basis of Savings
Chilled Water	Size Coils for 50 degF CHW, and Use Elevated CHW temp.	NEW or RETROFIT	Prevent simultaneous dehumidification / humidification
	Variable Speed Fans	NEW	Reduce parasitic fan heat losses.
Air-Cooled DX	Generous Sizing of Air-Cooled Condensers	NEW	Improve heat transfer and reduce approach temperature, improved refrigeration cycle.
	Keep Condenser Surfaces Clean	NEW or RETROFIT	Reduce head pressure for improved refrigeration cycle.
	Adjust Head Pressure Regulation Devices	NEW or RETROFIT	Prevent unintended false-loading of the refrigeration equipment in warm weather.
	Maintain Outdoor Equipment Spacing.	NEW	Prevent air recirculation, keep head pressure low for improved refrigeration cycle.
	Evaporative Pre-Cooling	NEW or RETROFIT	Reduce head pressure for improved refrigeration cycle. Reduce demand.
Water-Cooled DX	Select Equipment to Operate at Reduced Condenser Water Temp.	NEW	Reduce head pressure for improved refrigeration cycle.
	Use a Fluid Cooler instead of a Dry Cooler	NEW or RETROFIT	Reduce head pressure for improved refrigeration cycle.
	Adjust Water Regulation Valves	NEW or RETROFIT	Reduce head pressure for improved refrigeration cycle.
	Use Auxiliary Cooling Coils with a Fluid Cooler.	NEW or RETROFIT	Evaporative cooling reduces load on the refrigeration system, by pre-cooling or allowing the compressors to stop.
	Use Auxiliary Cooling Coils and Link to the Central Chilled Water System.	NEW or RETROFIT	Higher efficiency kW/ton at central cooling system compared to computer cooling equipment saves energy and Reduced run time of computer cooling compressors extends equipment life.
Common Items	Raise the Room Temperature	NEW or RETROFIT	Reduce thermodynamic lift by raising the refrigeration cycle low pressure.

Room Energy." Considering only the cooling equipment efficiency and the heat load it cools, the proportions of total energy that the cooling system consumes is shown.

Lighting: Higher efficiency lighting and occupancy sensor for unoccupied times.
Envelope: Locate away from envelope effects, especially glazing.
Maintenance: Calibrate every two years.
If a chiller is used: See "Chiller" this chapter.
- Water-side economizer

Sub-meter: If combined with other building uses

Table 3-1 (*Continued*). Computer Data Center Energy Savings Opportunities

System Type	Measure	New or Retrofit	Basis of Savings
Common Items Cont'd	Reduce Overhead Lighting Power.	NEW or RETROFIT	Reduce cooling heat load from lighting.
	Don't Provide Heaters in the Cooling Units	NEW	Savings in electrical infrastructure, including generator.
	Lower room humidity setting to 30% if possible.	NEW or RETROFIT	Prevent simultaneous dehumidification / humidification
	Use evaporative or ultrasonic humidification instead of infrared or resistance heat.	NEW or RETROFIT	More efficient technology for humidifying Part of cooling load displaced by adiabatic cooling
	Oversize Filters	NEW or RETROFIT	Reduce air flow resistance, in turn reducing fan motor kW and parasitic fan heat loss if fan speed is adjusted down.
	Premium Efficiency Fan Motors	NEW or RETROFIT	Reduced fan motor kW and parasitic fan heat loss.
	Raised Floor System Insulation and Air / Vapor Barrier	NEW	Reduced thermal loss to adjacent floor below. Reduced air and moisture losses by proper sealing of the plenum floor and walls, as well as plumbing and pipe penetrations.

EDUCATION—COLLEGES AND UNIVERSITIES

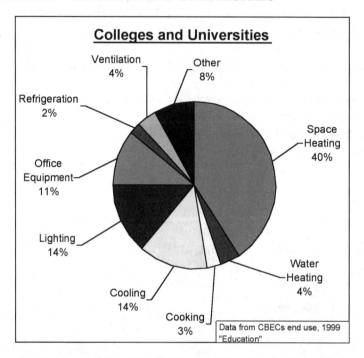

Figure 3-3. Energy End Use Distribution for Education/Colleges/Universities

Primary Energy Use Sources
- Heating and cooling
- Lighting
- Office equipment

Controls
- Scheduled start-stop of HVAC equipment and lighting.
- Set space temperatures to 75 cooling/70 heating.
- 5 degree deadband between heating and cooling operations.
- Eliminate simultaneous heating and cooling.
- Set space temperatures during unoccupied times up to 85/set back to 60 at night.
- Lock out cooling below 55 degrees and use only economizer.
- Extend the economizer cooling function, from 55 deg F to 65 deg F, if outdoor dew point levels are at or below indoor dew point levels—i.e. if not too humid outdoors.

- Optimum start and stop of primary cooling and heating equipment.
- Demand controlled ventilation for variable occupancy areas such as classrooms and lecture halls.
- For all roof openings to hoods and equipment not active in winter, dampers should be tightly closed during heating operation and when roof equipment is off.
- Demand limiting of central cooling equipment, and for VFD fans and pumps, to 85% capacity during peak demand periods.
- VAV—avoid supply air reset except in winter.
- VAV—avoid supply air reset that is based on return temperature.
- VAV—reduce supply duct static pressure.
- VAV—verify VAV box minimums are appropriately low and that the "heating minimum CFM" is no higher than the cooling minimum CFM.
- VAV—Classroom and lecture hall occupancy sensor to shut off HVAC systems when unoccupied for over 15 minutes. Perimeter rooms VAV HVAC goes to minimum air flow, interior rooms fully closed off.

Maintenance
- Calibrate controls every two years.
- Annual maintenance on all heat transfer surfaces, including good quality filters, cleaning coils, cleaning tubes.
- Use cog belts instead of standard V-belts for large motors.
- Seal return air plenums against any leakage to outside. Verify plenum temperature is within 2 degrees of room temperature in summer and winter design days.

Low Cost/No Cost
- Programmable thermostat for DX HVAC equipment.
- Install solar screening over large skylights and sun rooms.
- Occupancy sensors for lighting in classrooms, lecture hall, and labs.
- Photo cell control of parking lot and exterior lighting.
- Global control for "computer monitors off" after 15 minutes of inactivity, instead of screen savers.

Retrofit—Or Upgrade At Normal Replacement
Heating and Cooling
- Higher efficiency heating and cooling equipment.

- Convert constant volume HVAC to VAV.
- VFDs replace inlet vanes on air handlers.
- Evaporative pre-cooling on large air cooled package rooftop units.
- Anti-stratification fans for high bay, heated areas to move warm air to the floor.
- Dedicated cooling system for 24x7 needs in small areas, to allow the main building HVAC system to shut off at night.
- Reduce outside air to proper quantities if excessive.
- Increase roof insulation thickness to current energy code level as part of roof replacement.

Lighting
- Higher efficiency lighting.
- Replace incandescent lights with more efficient technology.
- Reduce excessive light levels by de-lamping.
- Combine de-lamping and reflectors to maintain light levels.

May Only Be Viable During New Construction
- Light color exterior walls.
- Light color high emissivity roof for low rise buildings.
- Passive shading elements for large glass areas.
- High performance low-E glazing.
- Exhaust heat recovery for high ventilation areas such as labs.
- Replace flat filters with angled filters.
- Circuiting of lights to allow first 10 feet inboard from the perimeter to be turned off during bright outdoor hours.
- Daylight lighting design.

Dormitories: See "Apartments" this chapter
Cafeteria: See "Food Service" this chapter
If a boiler is used: See "Boilers" this chapter
If a chiller is used: See "Chillers" this chapter
If the building is over four stories in height: See "High Rise" this chapter

EDUCATION—SCHOOLS K-12

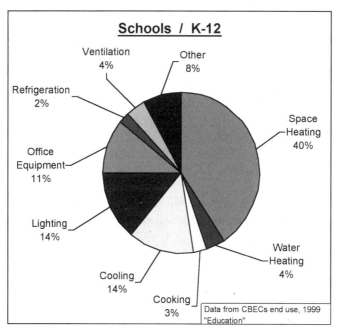

Figure 3-4. Energy End Use Distribution for Education—Schools K-12

Primary Energy Use Sources
- Heating and cooling
- Lighting
- Office equipment

Controls
- Scheduled start-stop of HVAC equipment and lighting.
- Set space temperatures to 75 cooling/70 heating.
- Set space temperatures during unoccupied times up to 85/set back to 60 at night.
- Lock out cooling below 55 degrees and use only economizer.
- Lock out heating above 65 deg F.
- Extend the economizer cooling function, from 55 deg F to 60 deg F, if outdoor dew point levels are at or below indoor dew point levels—i.e. if not too humid outdoors.
- 5 degree deadband between heating and cooling operations.

- Eliminate simultaneous heating and cooling.
- Unit ventilators use ASHRAE Control Cycle 2.
- For all roof openings to hoods and equipment not active in winter, dampers should be tightly closed during heating operation and when roof equipment is off.
- Demand controlled ventilation in assembly areas and other large controllable variable occupancy areas.
- VAV—avoid supply air reset except in winter.
- VAV—avoid supply air reset that is based on return temperature.
- VAV—reduce supply duct static pressure.
- VAV—verify VAV box minimums are appropriately low and that the "heating minimum CFM" is no higher than the cooling minimum CFM.
- VAV—Classroom occupancy sensor to shut off HVAC when unoccupied for over 15 minutes. Perimeter rooms VAV HVAC goes to minimum air flow, interior rooms fully closed off.

Maintenance
- Calibrate controls every two years.
- Annual maintenance on all heat transfer surfaces, including good quality filters, cleaning coils, cleaning tubes.
- Seal return air plenums against any leakage to outside. Verify plenum temperature is within 2 degrees of room temperature in summer and winter design days.

Low Cost/No Cost
- Programmable Thermostat for DX HVAC equipment.
- Occupancy sensors for lighting in classrooms and labs.
- Photo cell control of exterior lighting.

Retrofit—Or Upgrade At Normal Replacement
Heating and Cooling
- Higher efficiency heating and cooling equipment.
- Reduce outside air to proper quantities if excessive.
- Increase roof insulation thickness to current energy code level as part of roof replacement.

Lighting
- Higher efficiency lighting.

Energy Saving Opportunities by Business Type 33

- Replace incandescent lights with more efficient technology.
- Reduce excessive light levels by de-lamping.
- Combine de-lamping and reflectors to maintain light levels.

May Only Be Viable During New Construction
- Light color exterior walls.
- Light color high emissivity roof for low rise buildings.
- Passive shading elements for large glass areas.
- High performance low-E glazing.
- Exhaust heat recovery for high ventilation areas such as labs.
- Replace flat filters with angled filters.
- Circuiting of lights to allow first 10 feet inboard from the perimeter to be turned off during bright outdoor hours.
- Daylight lighting design.
- Ground source heat pumps.

Cafeteria: See "Food Service" this chapter
If a boiler is used: See "Boilers" this chapter
If a chiller is used: See "Chillers" this chapter

FOOD SALES—GROCERY STORES

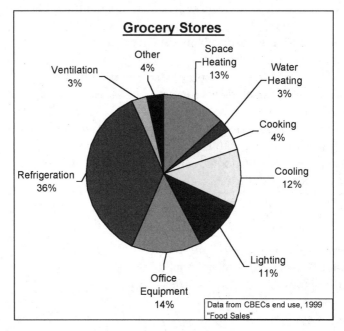

Figure 3-5. Energy End Use Distribution for Food Sales/Grocery Stores

Primary Energy Use Sources
- Refrigeration
- Office equipment
- Heating and cooling
- Lighting

Controls
- Demand controlled ventilation for variable occupancy
- Set back temperatures if not open 24 hours.

Maintenance
- Annual maintenance on all heat transfer surfaces, including good quality filters, cleaning coils, cleaning tubes.
- Calibrate HVAC controls every two years.
- Calibrate refrigeration controls every year.

Low Cost/No Cost
- Air curtains at customer entrance and shipping docks.

Retrofit—Or Upgrade At Normal Replacement
Refrigeration
- Higher efficiency refrigeration equipment.
- Convert air-cooled to water-cooled refrigeration equipment.

Heating and Cooling
- Higher efficiency heating and cooling equipment
- Evaporative pre-cooling on large air cooled package rooftop units in dry climates.
- Increase roof insulation thickness to current energy code level as part of roof replacement.

Lighting
- Higher efficiency lighting.

Other
- Power factor correction for groups of large motors.

May Only Be Viable During New Construction
- Heat recovery from refrigeration equipment for space heat or dehumidification.
- Light color high emissivity roof for low rise buildings.
- Daylight lighting design.

HEALTH CARE—HOSPITAL

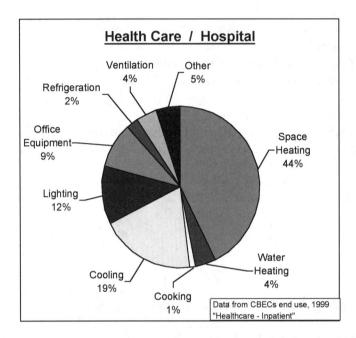

Figure 3-6. Energy End Use Distribution for Health Care—Hospital

Primary Energy Use Sources
- Heating and cooling
- Lighting
- Water heating

Controls
- Space temperatures set to 75 cooling/70 heating in live-in areas.
- Set space temperatures during unoccupied times up to 85/set back to 60 at night, for staff areas not continuously occupied.
- 5 degree deadband between heating and cooling operations.
- Eliminate simultaneous heating and cooling.
- Use outdoor economizer instead of mechanical cooling below 55 deg F, if prescribed space pressurization can be maintained.
- Extend the economizer cooling function, from 55 deg F to 60 deg F, if outdoor dew point levels are at or below indoor dew point levels— i.e. if not too humid outdoors.

- Constant volume HVAC—supply air reset from demand to reduce reheat.
- VAV—avoid supply air reset except in winter.
- VAV—avoid supply air reset that is based on return temperature.
- VAV—reduce supply duct static pressure.
- VAV—verify VAV box minimums are appropriately low and that the "heating minimum CFM" is no higher than the cooling minimum CFM.
- VAV—Meeting rooms and day-time office areas—occupancy sensor to shut off HVAC when unoccupied for over 15 minutes. Perimeter rooms VAV HVAC goes to minimum air flow, interior rooms fully closed off.
- Avoid humidification if possible.
- Lower domestic hot water temperature to 120 deg F for hand washing and bathing.
- Turn off domestic hot water re-circulation pumps at night to unoccupied areas of the building.

Maintenance
- Calibrate controls every two years.
- Annual maintenance on all heat transfer surfaces, including good quality filters, cleaning coils, cleaning tubes.
- Use cog belts instead of standard V-belts for large motors.

Low Cost/No Cost
- Occupancy sensors for lighting in meeting rooms and day-time office areas.
- Insulate bare hot piping for domestic water.
- Low flow faucets.

Retrofit—Or Upgrade At Normal Replacement
Heating and Cooling
- Higher efficiency heating and cooling equipment.
- Heat recovery from exhaust air, where exhaust is continuous.
- VFDs replace inlet vanes on air handlers.

Water Heating
- Low flow shower heads.
- Condensing domestic water heater.
- Stack damper for gas-fired domestic water heater.

Lighting
- Higher efficiency lighting.
- Replace incandescent lights with more efficient technology.
- Reduce excessive light levels by de-lamping.
- Combine de-lamping and reflectors to maintain light levels.

Other
- Power factor correction for groups of large motors.

May Only Be Viable During New Construction
- Replace flat filters with angled filters.
- Water-side economizer to make chilled water with just cooling towers, if air economizers are not used.
- VAV—supply and return box tracking instead of constant volume.
- Separate the water heating systems for 120, 140, and 160 degree hot water.

If a boiler is used: See "Boilers" this chapter
- Boiler stack gas heat recovery.

If a chiller is used: See "Chillers" this chapter

If the building is over four stories in height: See "High Rise" this chapter

---

HEALTH CARE—NON HOSPITAL

Primary Energy Use Sources
- Heating and cooling
- Cooking (live-in facilities)
- Lighting

Controls
- Scheduled start-stop of HVAC equipment and lighting.
- Set space temperatures to 75 cooling/70 heating. Avoid over-cooling.

Energy Saving Opportunities by Business Type 39

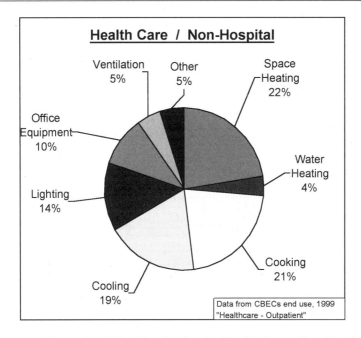

Figure 3-7. Energy End Use Distribution for Health Care—Non Hospital

- Set space temperatures during unoccupied times up to 85/set back to 60 at night.
- 5 degree deadband between heating and cooling operations.
- Eliminate simultaneous heating and cooling.
- Lock out cooling below 55 degrees and use only economizer.
- Lock out heating above 65 deg F.
- Extend the economizer cooling function, from 55 deg F to 60 deg F, if outdoor dew point levels are at or below indoor dew point levels—i.e. if not too humid outdoors.
- Optimum start and stop of primary cooling and heating equipment.
- Constant Volume HVAC—supply air reset from demand to reduce reheat.
- VAV—avoid supply air reset except in winter.
- VAV—avoid supply air reset that is based on return temperature.
- VAV—reduce supply duct static pressure.
- VAV—verify VAV box minimums are appropriately low and that the "heating minimum CFM" is no higher than the cooling minimum CFM.
- Avoid humidification if possible.

- Eliminate simultaneous humidification/dehumidification
- Demand limiting of central cooling equipment, and for VFD fans and pumps, to 85% capacity during peak demand periods.
- Demand controlled ventilation in office areas.

Maintenance
- Calibrate controls every two years.
- Annual maintenance on all heat transfer surfaces, including good quality filters, cleaning coils, cleaning tubes.
- Seal return air plenums against any leakage to outside. Verify plenum temperature is within 2 degrees of room temperature in summer and winter design days.

Low Cost/No Cost
- Programmable thermostat—use set up/set back feature at night unless serving sensitive equipment.
- Occupancy sensors for lighting in meeting rooms and restrooms.

Retrofit—Or Upgrade At Normal Replacement
Heating and Cooling
- Higher efficiency heating and cooling equipment.
- Dedicated small cooling system for 24x7 needs in small areas, to allow the main building HVAC system to shut off at night.
- Reduce outside air to proper quantities if excessive.
- Increase roof insulation thickness to current energy code level as part of roof replacement.

Lighting
- Higher efficiency lighting.
- Replace incandescent lights with more efficient technology.
- Reduce excessive light levels by de-lamping.
- Combine de-lamping and reflectors to maintain light levels.

May Only Be Viable During New Construction
- Light color high emissivity roof for low rise buildings.
- Replace flat filters with angled filters.

Kitchen: See "Food Service" this chapter

Energy Saving Opportunities by Business Type 41

LAUNDRIES—COMMERCIAL

Primary Energy Use Sources
- Water heating
- Heat for tumble drying
- Steam for pressing

Controls
- Turn off domestic hot water re-circulation pumps at night.

Maintenance
- Annual cleaning of heat transfer surfaces, including water heaters, heat recovery equipment.

Low Cost/No Cost
- Insulate bare hot piping.
- Reduce water temperature if possible.

Retrofit—Or Upgrade At Normal Replacement
Water Heating
- Heat recovery for waste water, to preheat cold water.
- Heat recovery for hot exhaust from dryers and roller irons, to preheat cold water.
- Condensing water heater.
- Stack damper for water heater.

Cooling
- Evaporative cooling.

May Only Be Viable During New Construction
- Washers that use less water.
- High speed extractors.

Boiler: See "Boilers" this chapter

LIBRARIES/MUSEUMS

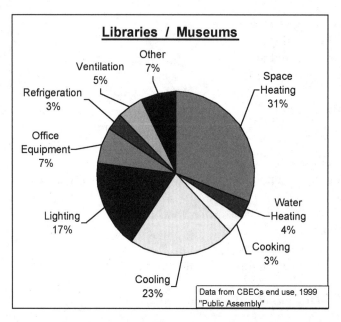

Figure 3-8. Energy End Use Distribution for Libraries/Museums

Primary Energy Use Sources
- Heating and cooling
- Lighting

Controls
- Scheduled start-stop of lighting even if HVAC must run continuously.
- Set space temperature controls to avoid over-cooling.
- 5 degree deadband between heating and cooling operations if possible.
- Eliminate simultaneous heating and cooling.
- If air economizer is not allowed for humidity stabilization, use water economizer or dry cooler below 55 degrees outside temperature.
- Lock out cooling below 50 degrees F.
- Lock out heating above 65 deg F unless dehumidification cycle is needed.
- Constant volume HVAC—supply air reset from demand to reduce reheat.

- VAV—avoid supply air reset except in winter.
- VAV—avoid supply air reset that is based on return temperature.
- VAV—reduce supply duct static pressure.
- VAV—verify VAV box minimums are appropriately low and that the "heating minimum CFM" is no higher than the cooling minimum CFM.
- Avoid humidification if possible.
- Eliminate simultaneous humidification/dehumidification.
- Demand controlled ventilation for unoccupied periods and variable occupancy.

Maintenance
- Annual maintenance on all heat transfer surfaces, including good quality filters, cleaning coils, cleaning tubes.
- Calibrate HVAC controls every two years.

Low Cost/No Cost
- Light sensors to harvest daylight near skylights and reading areas near glass.

Retrofit—Or Upgrade At Normal Replacement
Heating and Cooling
- Higher efficiency heating and cooling equipment.
- Install solar screening over large skylights and sun rooms.
- Install evaporative pre-cooling on air-cooled chillers in dry climates.

Lighting
- Higher efficiency lighting.
- Replace incandescent lights with more efficient technology.
- Reduce excessive light levels by de-lamping.

May Only Be Viable During New Construction
- Light color high emissivity roof for low rise buildings.

Boiler: See "Boilers" this chapter
Chiller: See "Chillers" this chapter

LODGING/HOTELS/MOTELS

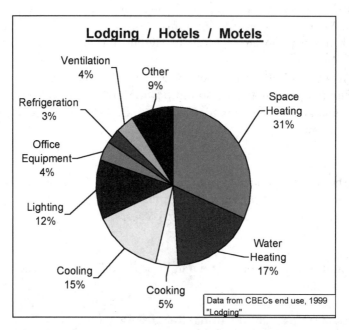

Figure 3-9. Energy End Use Distribution for Lodging/Hotels/Motels

Primary Energy Use Sources
- Heating and cooling
- Water heating
- Lighting

Controls
- Scheduled start-stop of HVAC equipment and lighting in staff areas and other areas without continuous guest access.
- Set space temperatures to 75 cooling/70 heating.
- Set space temperatures during unoccupied times up to 85/set back to 60 at night, for staff areas not continuously occupied and amenity areas without continuous guest access.
- Lock out cooling below 55 degrees and use only economizer (common and amenity areas).
- Lock out heating above 65 deg F (common and amenity areas).
- Extend the economizer cooling function, from 55 deg F to 60 deg F, if

Energy Saving Opportunities by Business Type 45

outdoor dew point levels are at or below indoor dew point levels—i.e. if not too humid outdoors.
- Eliminate overlapping heating and cooling in common area HVAC equipment by adjusting control settings to have a "deadband."
- Lower domestic hot water temperature to 120 deg F for hand washing and bathing, and 140 deg F for dish washing, if possible.
- Turn off domestic hot water re-circulation pumps at night.
- Guest occupancy controls to shut off lights and set back HVAC.

Larger Facilities
- Demand controlled ventilation for variable occupancy areas such as meeting rooms, ballrooms, and common areas.
- Avoid over-cooling large meeting rooms and ballrooms. Limit these space temperatures to 70 degrees.
- Demand limiting of central cooling equipment, and for VFD fans and pumps, to 85% capacity during peak demand periods.

Maintenance
- Calibrate controls every two years.
- Annual maintenance on all heat transfer surfaces, including good quality filters, cleaning coils, cleaning tubes.
- Use cog belts instead of standard V-belts for large motors.
- Seal return air plenums against any leakage to outside. Verify plenum temperature is within 2 degrees of room temperature in summer and winter design days.
- Repair or replace defective zone control valves.

Low Cost/No Cost
- Insulate bare hot piping for domestic water.
- Install solar screening over large skylights and sun rooms.
- Occupancy sensors for lighting in amenity areas and meeting rooms.
- Close PTAC "vent" function if corridor ventilation is adequate.

Retrofit—Or Upgrade At Normal Replacement
Heating and Cooling
- Higher efficiency heating and cooling equipment.
- Higher efficiency guest room package terminal unit air conditioners (PTACs).

- In mild climates, heat pump PTACs instead of electric heat.
- Increase roof insulation thickness to current energy code level as part of roof replacement.

Water Heating
- Low flow shower heads.
- Stack damper for gas-fired domestic water heater.
- Condensing domestic water heater.

Lighting
- Higher efficiency lighting.
- Replace incandescent lights with more efficient technology.
- Reduce excessive light levels by de-lamping.
- Combine de-lamping and reflectors to maintain light levels.

Larger Facilities
- Install solar screening over large skylights and sun rooms.
- Convert constant volume HVAC to VAV.
- VFDs replace inlet vanes on air handlers.
- Power factor correction for groups of large motors.
- Dedicated cooling system for 24x7 needs in small areas, to allow the main building HVAC system to shut off at night.
- Reduce outside air to proper quantities if excessive.
- Low water volume/high speed spin wash machines (Energy Star).

May Only Be Viable During New Construction
- Envelope construction in guest areas to match local residential energy code for insulation.
- Overhangs or other exterior shading at windows.
- Light color high emissivity roof for low rise buildings.
- Toilet exhaust on demand (light switch) instead of continuous.
- Low volume bathtubs.
- Separate the water heating systems for 120, 140, and 160 degree hot water.
- Gas-fired PTACs instead of electric resistance.
- Ground source heat pumps.

Kitchen: See "Food Service" this chapter
Pool: See "Pools" this chapter

Energy Saving Opportunities by Business Type 47

If a boiler is used: See "Boilers" this chapter
If a chiller is used: See "Chillers" this chapter
If the building is over four stories in height: See "High Rise" this chapter

OFFICE BUILDINGS

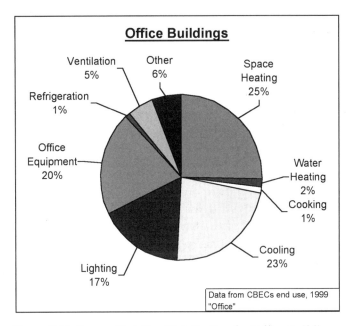

Figure 3-10. Energy End Use Distribution for Office Buildings

Primary Energy Use Sources
- Heating and cooling
- Lighting
- Office equipment

Controls
- Scheduled start-stop of HVAC equipment and lighting.
- Set space temperatures to 75 cooling/70 heating.

- Set space temperatures during unoccupied times up to 85/set back to 60 at night.
- 5 degree deadband between heating and cooling operations.
- Eliminate simultaneous heating and cooling.
- Lock out cooling below 55 degrees and use only economizer.
- Lock out heating above 65 deg F.
- Extend the economizer cooling function, from 55 deg F to 60 deg F, if outdoor dew point levels are at or below indoor dew point levels—i.e. if not too humid outdoors.
- VAV—avoid supply air reset except in winter.
- VAV—avoid supply air reset that is based on return temperature.
- VAV—reduce supply duct static pressure.
- VAV—Verify VAV box minimums are appropriately low and that the "heating minimum CFM" is no higher than the cooling minimum CFM.

Larger Facilities
- Optimum start and stop of primary cooling and heating equipment.
- Stage electric heating to prevent setting seasonal demand and invoking ratchet charges.
- VAV—Conference room occupancy sensor to shut off HVAC when unoccupied for over 15 minutes. Perimeter rooms VAV HVAC goes to minimum air flow, interior rooms fully closed off.
- Demand controlled ventilation for variable occupancy areas such as conference rooms and open plan office areas.
- Demand limiting of central cooling equipment, and for VFD fans and pumps, to 85% capacity during peak demand periods.

Maintenance
- Calibrate controls every two years.
- Annual maintenance on all heat transfer surfaces, including good quality filters, cleaning coils, cleaning tubes.
- Use cog belts instead of standard V-belts for large motors.
- Seal return air plenums against any leakage to outside. Verify plenum temperature is within 2 degrees of room temperature in summer and winter design days.

Low Cost/No Cost
- Programmable thermostat for package A/C units—set up to 85 deg/ set back to 60 deg at night.

Energy Saving Opportunities by Business Type   49

- Install solar screening over large skylights and sun rooms.
- Occupancy sensors for lighting in conference rooms and meeting rooms.
- Global control for "computer monitors off" after 15 minutes of inactivity, instead of screen savers.

Retrofit—Or Upgrade At Normal Replacement
Heating and Cooling
- Higher efficiency heating and cooling equipment.
- Convert constant volume HVAC to VAV.
- VFDs replace inlet vanes on air handlers.
- Reduce outside air to proper quantities if excessive.
- Increase roof insulation thickness to current energy code level as part of roof replacement.

Lighting
- Higher efficiency lighting.
- Replace incandescent lights with more efficient technology.
- Reduce excessive light levels by de-lamping.
- Combine de-lamping and reflectors to maintain light levels.

Larger Facilities
- Evaporative pre-cooling on large air cooled package rooftop units in dry climates.
- Power factor correction for groups of large motors.
- Dedicated cooling system for 24x7 needs in small areas, to allow the main building HVAC system to shut off at night.
- Window film or light colored interior shades for buildings with large amounts of glass.

May Only Be Viable During New Construction
- High performance low-E glazing
- Overhangs or exterior shading for glazing.
- Light color exterior walls.
- Light color high emissivity roof for low rise buildings.
- Circuiting of lights to allow first 10 feet inboard from the perimeter to be turned off during bright outdoor hours.

Data center: See "Data Centers" this chapter
If a boiler is used: See "Boilers" this chapter

If a chiller is used: see "Chillers" this chapter
If the building is over four stories in height: See "High Rise" this chapter

FOOD SERVICE/RESTAURANTS

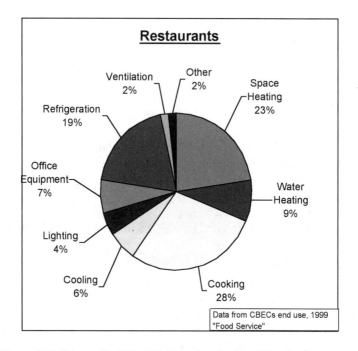

Figure 3-11. Energy End Use Distribution for Food Service/Restaurants

Primary Energy Use Sources
- Cooking
- Heating and cooling
- Refrigeration
- Water heating

Kitchen and Equipment
- Run hoods only when needed.

Energy Saving Opportunities by Business Type 51

- Control multiple hoods independently and provide variable demand-based make-up air.
- Turn down cooking equipment to standby temperature when not in use.
- Provide controls so the food warming lamps only operate when there is food there to be warmed, instead of always on.
- Limit kitchen cooling to 75 degrees.
- Limit freezer temperature to zero deg F—put ice cream in a separate small chest freezer.
- Limit cooler temperature to 37 deg F.
- Limit hood make-up air tempering to 50 degrees heating.
- Disable the heating for the HVAC units that serve the kitchen area, or lower the setting to only come on below 60 degrees. This is a "cooling only" application and should not require heating.
- Install interlock for the dishwasher hood exhauster so that it only runs when the dishwasher runs.
- Gaskets and door sweeps on walk-in cooler and freezer doors.
- Hanging plastic strips at openings into walk-in coolers and freezers.
- Locate condensing units to reject heat outside, for coolers, freezers, ice makers.
- Use chemical sanitizer rinse instead of 180 deg F booster heater.
- Variable flow heat hoods (Class 1)
- UL listed hoods for reduced air flows.
- Direct fired hood make-up heating.
- Separate hot and cold equipment used for food storage and preparation.
- Air balance to assure proper hood capture of hot cooking fumes.
- Walk-in freezer and cooler combined so entrance to freezer is from the walk-in cooler.
- On-demand cooking equipment such that energy use closely tracks production, and will allow proportionally reduced energy cost during slow times.

Controls
- Scheduled start-stop of HVAC equipment and lighting.
- Set space temperatures to 75 cooling/70 heating.
- Set space temperatures during unoccupied times up to 85/set back to 60 at night.
- Lock out cooling below 55 degrees and use only economizer.

- Lock out heating above 65 deg F.
- Lower domestic hot water temperature to 120 deg F for handwashing, and 140 deg F for dish washing, if possible.
- Turn off domestic hot water re-circulation pumps at night.
- (Dining areas) demand controlled ventilation unless tied to kitchen exhaust/make-up system.

Maintenance
- Calibrate controls every two years.
- Each three months or less: maintenance on all heat transfer surfaces, including good quality filters, cleaning coils, cleaning tubes. Include refrigeration equipment, coolers, and freezers.

Low Cost/No Cost
- Programmable thermostat to prevent HVAC equipment from running continuously at night.
- Install solar screening over large skylights and sun rooms.
- Insulate bare hot piping for domestic water.

Retrofit—Or Upgrade At Normal Replacement
Heating and Cooling
- Add A/C economizer for kitchen area, if none exists.
- Higher efficiency cooling equipment.
- Down-size HVAC cooling equipment if over-sized.

Water Heating
- Stack damper for gas-fired domestic water heater.
- Condensing domestic water heater.

Energy Saving Opportunities by Business Type

RETAIL/SALES

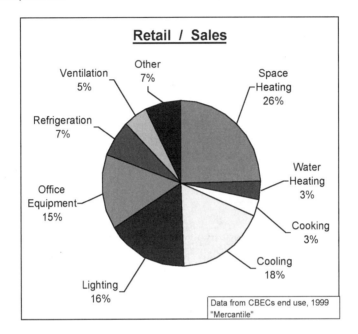

Figure 3-12. Energy End Use Distribution for Retail/Sales

Primary Energy Use Sources
- Heating and cooling
- Lighting
- Office equipment

Controls
- Scheduled start-stop of HVAC equipment and lighting.
- Set space temperatures to 75 cooling/70 heating.
- Set space temperatures during unoccupied times up to 85/set back to 60 at night.
- Lock out cooling below 55 degrees and use only economizer.
- Lock out heating above 65 deg F.
- Extend the economizer cooling function, from 55 deg F to 605 deg F, if outdoor dew point levels are at or below indoor dew point levels— i.e. if not too humid outdoors.
- Maintain a 5 deg F deadband between heating and cooling for all HVAC equipment.

- Turn off display lighting except during customer times.

Larger Facilities
- Optimum start and stop of primary cooling and heating equipment.
- Demand controlled ventilation for variable occupancy times.
- Photo cell control of parking lot lighting.
- For all roof openings to hoods and equipment not active in winter, dampers should be tightly closed during heating operation and when roof equipment is off.
- If demand metered: demand limiting of central cooling equipment, and for VFD fans and pumps, to 85% capacity during peak demand periods.

Maintenance
- Calibrate controls every two years.
- Annual maintenance on all heat transfer surfaces, including good quality filters, cleaning coils, cleaning tubes.

Larger Facilities
- Use cog belts instead of standard V-belts for large motors.
- Seal return air plenums against any leakage to outside. Verify plenum temperature is within 2 degrees of room temperature in summer and winter design days.

Low Cost/No Cost
- Programmable thermostat for package A/C units—set up to 85 deg/ set back to 60 deg at night.
- Occupancy sensors for lighting in break rooms.
- Install solar screening over large skylights.

Retrofit—Or Upgrade At Normal Replacement
Heating and Cooling
- Higher efficiency heating and cooling equipment.
- HVAC air economizer, lock out mechanical cooling below 55 deg F.
- Increase roof insulation thickness to current energy code level as part of roof replacement.

Lighting
- Higher efficiency lighting.

- Replace incandescent lighting with more efficient technology.
- Reduce excessive light levels by de-lamping.
- Combine de-lamping and reflectors to maintain light levels.

Larger Facilities
- High bay lighting retrofit, HID to fluorescent.
- Evaporative pre-cooling on large air cooled package rooftop units.
- Convert constant volume HVAC to VAV.
- VFDs replace inlet vanes on air handlers.
- Power factor correction for groups of large motors.
- Reduce outside air to proper quantities if excessive.

May Only Be Viable During New Construction
- Overhang over storefront window.
- HVAC variable volume air systems.
- Light color high emissivity roof for low rise buildings.
- Daylight lighting design.

If a boiler is used: see "Boilers" this chapter
If a chiller is used: see "Chillers" this chapter
If the building is over four stories in height: see "High Rise" this chapter

WAREHOUSES

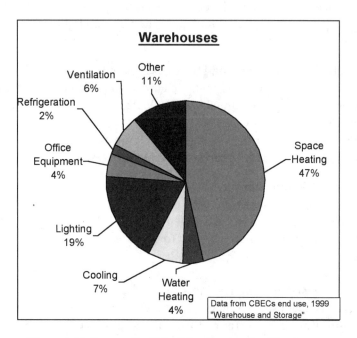

Figure 3-13. Energy End Use Distribution for Warehouses

Primary Energy Use Sources
- Heating
- Lighting

Controls
- Occupancy sensors to turn off lights in unmanned areas and during unoccupied times.
- Set space temperatures to 80 cooling/60 heating unless product storage requires closer temperature control.
- Set space temperatures during unoccupied times up to 90/set back to 50 at night unless product storage requires closer temperature control.
- For all roof openings to hoods and equipment not active in winter, dampers should be tightly closed during heating operation and when roof equipment is off.

Energy Saving Opportunities by Business Type

Maintenance
- Annual maintenance on all heat transfer surfaces, including good quality filters, cleaning coils.

Low Cost/No Cost
- Interlock heating and cooling at loading docks to stop when roll-up doors are opened.

Refrigerated Warehouse
- Hanging plastic strips at openings into refrigerated areas.
- Air curtains at openings into refrigerated areas.

Retrofit—Or Upgrade At Normal Replacement:
Heating and Cooling
- Higher efficiency heating equipment.
- Gas-fired radiant heating in lieu of space heating.
- Anti-stratification fans for high bay, heated areas to move warm air to the floor.
- Increase roof insulation thickness to current energy code level as part of roof replacement, for buildings that are heated or cooled.

Lighting
- High bay lighting retrofit HID to fluorescent.
- Higher efficiency lighting.
- Motion sensors for overhead lights, except re-strike time makes this impractical for HID lighting.
- Reduce excessive light levels by de-lamping.
- Combine de-lamping and reflectors to maintain light levels.

Refrigerated Warehouse
- Higher efficiency lighting to reduce refrigeration load.
- Convert air-cooled to water-cooled refrigeration.
- Increase roof insulation thickness over refrigerated areas.

May Only Be Viable During New Construction
- Improved insulation for buildings intended to be heated or cooled

Refrigerated Warehouse
- Improve envelope insulation

- Light color exterior walls for refrigerated areas.
- Light color high emissivity roof for refrigerated areas.
- Daylight lighting design.

BOILERS

- Controls to lock out boiler above 60 deg F.
- Reset boiler temperature from outside air.
- Annual verification that hydronic heating and cooling automatic control valves close tightly and prevent any internal leak-by.
- Modular boilers to reduce standby losses, upon replacement.
- Stack dampers for single boilers that cycle frequently during part load.
- Annual efficiency checking. Take corrective action if efficiency is found to be less than 95% of new equipment values.
- Automatic boiler isolation valves if piping allows hot water through an "off" standby boiler.
- Insulate bare heating water piping.
- Maintain proper water treatment and ensure that normal make-up does not cause dilution and scaling.
- Convert constant flow heating water to variable flow for larger systems with high annual run hours.
- Separate domestic water heating equipment, instead of using heating boiler during summer for this purpose.
- Insulate boiler surface areas and access panels that have no casing insulation.
- Reduce excess air for burner to 30% or less.

CHILLERS

- Higher efficiency chiller.
- Vary auxiliary pump flows in proportion to cooling load instead of constant flow.

- Lower condenser water temperature for water-cooled chiller.
- Higher efficiency/capacity cooling tower upon replacement. Suggested criteria for replacement cooling tower is a maximum design approach of 7 degrees to design wet bulb conditions and a fan power budget of no more than 0.05 kW/ton.
- VFDs for cooling tower fans.
- Annual condenser tube cleaning for water-cooled chiller.
- Controls to lock out the chiller below 50 deg F.
- Evaporative pre-cooling for air-cooled chillers.
- Raise chilled water temperature if possible, but use caution when dehumidification is needed.
- Convert constant flow chilled water to variable flow for larger systems with high annual run hours.
- Insulate bare chilled water piping.

HIGH RISE

Envelope
- Entry door vestibule or revolving door.
- Seal around punch windows.
- Seal vertical shafts.
- Seal air plenums.
- Motorized damper to close elevator shaft except in fire mode, if allowed by local building regulations.

POOLS

Primary Energy Use Sources
- Pool water heating
- Pool air heating and dehumidifying
- Shower water heating

Controls
- Maintain air temperature within 2 degrees above water temperature.
- Maintain air humidity above between 50-60%, do not over dry the air through excessive ventilation or other means.
- Reduce pool water temperature overall.
- Relax pool water temperature, air temperature, and humidity requirements during unoccupied times. Allow water to cool a few degrees at night if possible.
- Reduce air flow rates and air exchange rates in unoccupied periods.
- For all roof openings to hoods and equipment not active in winter, dampers should be tightly closed during heating operation and when roof equipment is off.

Maintenance
- Calibrate controls every two years.
- Annual efficiency checks for gas fired heaters.
- Annual maintenance on all heat transfer surfaces, including good quality filters, cleaning coils, cleaning tubes.
- Look for condensation on interior walls during winter that indicate poor insulation.

Low Cost/No Cost
- Drain outdoor pools in winter instead of heating.
- Minimize sand filter backwash, usually once per week is sufficient.

Retrofit—Or Upgrade at Normal Replacement
Pool Water Heating
- Condensing water heater.
- Filtration instead sand filters, to eliminate backwash and make-up water heating.
- Pool covers.

Pool Air Heating and Dehumidifying
- Reduce outside air to proper quantities if excessive.

Shower Water Heating
- Low flow shower heads.

Energy Saving Opportunities by Business Type 61

May Only Be Viable During New Construction
- Replace mechanical cooling system with ventilation system in climates with acceptable summer humidity levels.
- For indoor pools, add a heating system for the surrounding air if there isn't one, to eliminate the pool from heating it.

ICE RINKS

Primary Energy Use Sources
- Refrigeration
- Space Heating
- Lighting

Controls
- Stage brine pumps proportionally with load.
- Control ice sheet temperature no lower than necessary.
- Control brine temperature no lower than necessary.
- Control condenser water temperature as low as possible to reduce head pressure.
- Turn off heaters in ice areas in unoccupied times.
- Closely monitor both temperature and humidity of outside air and only use outside air beyond minimums when beneficial.

Maintenance
- Calibrate controls every two years.
- Annual efficiency checks for gas fired heaters.
- Annual maintenance on all heat transfer surfaces, including good quality filters, cleaning coils, cleaning tubes.

Low Cost/No Cost
- Schedule ice-building times to use off-peak utility rates.
- Hanging barriers to keep warmed air for spectators from heating the ceiling over the ice slab.
- Window shades for glazing at ice slabs.
- Re-direct warm air heating supply to spectators but not at or over the ice sheet.

Retrofit—Or Upgrade at Normal Replacement
Refrigeration
- Replace conventional expansion valve with electronic expansion valve to allow colder condenser water, lower head pressures.
- Higher efficiency refrigeration equipment.
- Higher efficiency cooling tower with ample capacity (smaller fan motor)
- Variable speed brine pumping.
- Heat recovery from refrigeration system for space heating or for ice melting.
- Light colored roof over ice sheet.
- Increased wall and roof insulation around ice sheet.
- Increased insulation under ice sheet.
- Low-emissivity paint in interior roof over ice sheet.
- Heavy tint or frittered glazing near ice sheet.
- Switch to ethylene glycol if propylene glycol is being used.
- High efficiency lighting over ice sheet.
- Barriers to keep heated air over spectators from heating the ice sheet.

Space Heating
- Higher efficiency heating equipment
- Radiant heat for spectators instead of air heating.

Lighting
- Higher efficiency lighting has double effect by reduced load on the ice sheet.

Chapter 4
Manufacturing

GENERAL

Manufacturing is often called a business sector, but is really a collection of dozens of other sectors. This book does not attempt to discuss each manufacturing sub-sector in detail, however some generalities are provided that may be useful.

Some manufacturing processes are very energy intensive, while others are not. Therefore, the effect of outdoor temperatures and climate can vary from substantial to insignificant. More often, the energy use of a manufacturing facility is driven by the process. In these facilities, the monthly load profile variations are strongly dependent on production volumes.

Stand By Losses: Manufacturing tool energy use measurements (amps) are useful to determine how well the tool energy use will turn down along with the process; i.e., what fraction of the full load production energy use is spent while idling. Many modern tools turn down to 10% or less in a "sleep mode," while many older tools consume half of the full load energy just sitting there. The higher energy use in idle mode has a pronounced effect on plant profitability during periods of reduced output.

PROCESS ANALYSIS

As part of the audit, use a manufacturing process flow chart and observe the process in each sequential step, and identifying the energy input in each step. **See Figure 4-1.** The flow chart can either be one provided by the manufacturer, or can be created on a notepad from interviews with plant personnel. This approach can often provide insight into savings opportunities. If possible, identifying the energy use for each step is very helpful since it draws attention to the higher energy use areas. This step may be very laborious but is the only way to truly understand the energy picture. Including sufficient detail to account for all the energy use is the goal. Then, the constituent parts identify an end-use pie diagram, showing which processes use the larger percentage of the total energy amount.

Use the process flow diagram to indicate where energy and other utilities are input. This step can sometimes reveal process steps that can be linked for energy benefit.

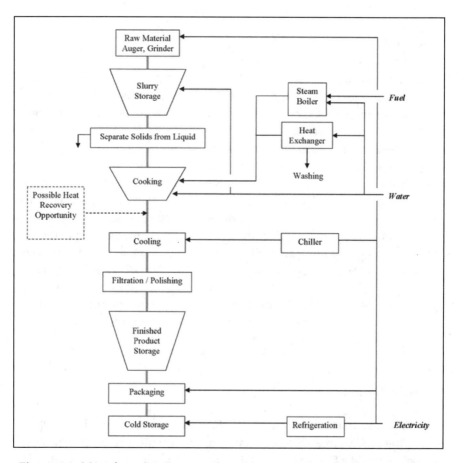

Figure 4-1. Manufacturing Process Flow Diagram with Utility Inputs Added (Example Diagram)

In most plants, 10-15% of the equipment in use accounts for the majority of a facility's energy consumption. Source: "Save Energy Now," Industrial Technologies Program (ITP), 2006, US DOE Office of Energy Efficiency and Renewable Energy.

For effective communication, it is important to understand that the

EXAMPLE MANUFACTURING PROCESS FLOW CHART

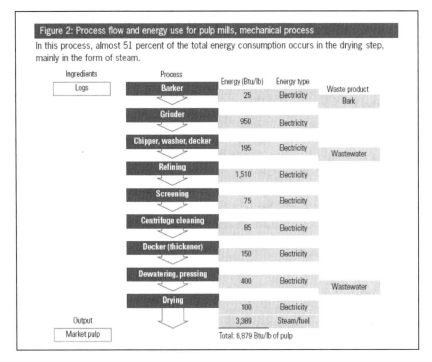

Figure 4-2. Manufacturing Process Flow Chart with Energy Use Identified.
Source: E Source, 2004; data from H.L. Brown

manufacturing customer is often more in tune with increasing production rate in their facility than reducing energy use.

DESIRE FOR ENERGY USE TO FOLLOW PRODUCTION RATES

A very important first step is to equate percent energy use to percent production for the same period. Graphs are very helpful for this. If energy use changes do not closely follow production rate changes, this probably indicates high standby losses such as ovens left on or machinery that will not fully unload while idling. **See Chapter 2—Analyzing Energy Use Graphs "Business Volume (Production Rates)"**

PRIMARY ENERGY USE SOURCES

- Most energy use is directly proportional to production.
- End use breakdown depends entirely on the process.
- Choice of measures depends upon the process.

PRODUCTION SCHEDULING

- For companies running multiple shifts, evaluate operations that are on-peak and off-peak. It may be possible to modify the current production sequence to use high energy consuming equipment during off-peak times. This may require intermediate storage of materials.
- Where shift pay differentials exist, labor costs will usually outweigh energy savings from adding another shift or moving to night shift to leverage off peak rates.

MAINTENANCE

- Annual maintenance on all heat transfer surfaces, including good quality filters, cleaning coils, cleaning tubes. Increase frequency of cleaning where process causes rapid fouling of heat exchange surfaces.
- Locate and repair leaking oven door seals.
- Use cog belts instead of standard V-belts for large motors.
- Survey and log fired heat exchanger process flue temperatures and other indicators of fouling, for predictive maintenance. Use heat exchange "approach" as a benchmark and indicator of waste.
- Maintain proper boiler water treatment and ensure that normal make-up does not cause dilution and scaling.

CONTROLS

- Monitor and track energy use as a function of production, to establish benchmark data and cost influence on final product. Identify energy use rate in terms of Btu per pound, ton, or cubic foot of product, Btu per quantity of product, etc. Use this benchmark to measure the consistency of energy use "per product" and to measure results of

Manufacturing

process changes. Note: the approach of manufacturing ECMs may be to reduce the energy cost per product output; or to increase production output without a proportional increase in energy input.
- Control process equipment to make energy use track production use; increasing and decreasing proportionally with production rates.
- Turn off unused equipment.
- Determine the cost impact of just-in-time manufacturing.
- Eliminate simultaneous heating and cooling.
- Use controls to optimize processes. Provide enough, but just enough, to do the job properly.
- Lower settings for high bay space heaters.
- For precision temperature control processes using both heating and cooling medium, widen control deadband if possible to reduce the overlap.
- For all roof openings to hoods and equipment not active in winter, automatic dampers should be tightly closed during heating operation and when roof equipment is off.
- Provide process ventilation and make-up only when needed, instead of continuously.
- Utilize optimized control settings that allow adaptation for production rates, and seasonal effects, instead of one-size-fits-all fixed set points. Fixed set points almost always translate into energy use increase.

SOME COMMON ECMS FOR MANUFACTURING

Low Cost / No Cost
- Turn off equipment that is not in use for over a half hour.
- Lower stand-by temperatures for ovens, mold machines, etc.
- Schedule work such that electrical load is leveled, to avoid high utility costs associated with poor load factor (average demand divided by maximum demand, in percent).

Retrofit—Or Upgrade at Normal Replacement
- Switch from air conditioning to evaporative cooling.
- Switch from air conditioning to spot cooling.
- Switch from mechanically cooled process cooling water to a fluid cooler and evaporative cooling.

- Heat recovery for adjacent and coincident heating and cooling processes. Waste heat can serve for process use or for space heating. Large sources of make-up air are candidates for preheating.
- When heating and cooling operations are weather-independent, heat recovery opportunities are improved because available hours per year are increased.
- Evaporative pre-cooling on large air cooled chillers and air package rooftop units.
- Boiler stack gas heat recovery.
- Rinse-water or wash-water effluent heat recovery to the concurrent make-up water stream.
- High bay lighting retrofit from HID to fluorescent.
- Roof insulation for heated buildings, as part of roof replacement, for heated or cooled buildings.
- Anti-stratification fans for high bay, heated areas to move warm air to the floor.
- Lids, floating balls, or other means to reduce evaporation from heated tanks.
- Power factor correction for groups of large motors, inverters, and welders.
- Higher efficiency motors.
- Higher efficiency cooling equipment (HVAC or process cooling).
- Insulate bare hot piping for process, including hot water, steam and condensate, vats, heat exchangers, tanks, furnace casings, boiler casings, etc.
- Process cooling equipment condensers vent to outdoors instead of inside the facility.
- Process ovens retrofitted with automatic doors or other means to minimize heat release.
- Reduce excessive light levels by de-lamping.
- Combine de-lamping and reflectors to maintain light levels.
- Variable flow heat hood exhaust. If the exhaust captures just heat, vary the exhaust flow rate to maintain a leaving temperature 30 degrees higher than surrounding indoor ambient air temperature so the system always acts like a spot exhauster and reduces the volume make-up air.
- Large steam uses after pressure reducing valves may be an opportunity for a separate low pressure steam system.
- Very large steam uses after pressure reducing valves may be an op-

Manufacturing 69

portunity for a steam work extraction element (turbine driven device or cogeneration).
- Combine processes where possible, to reduce "new energy" input. Examples are using rejected heat from an air compressor for heating, rejected from one process to regenerate a desiccant bed, rejected heat from a cooling operation to a concurrent heating need.

If There are Office Areas
See "Office Buildings" this chapter.

Process Boilers, Furnaces, Ovens, and Other Combustion Heating Systems
- Monitor stack gas temperature by routine measurement. High or increasing heat exchange approach temperatures often indicate fouling and need for cleaning.
- Monitor combustion efficiency by routine measurement. Verify air-fuel ratio is accurately controlled at all loads (100%, 50%, 25% firing rate). Reduce excess air.
- Turn equipment off if there are long periods of idle time between batches
- Reduce operating temperature if possible. Provide just enough heating. Optimal temperature setting may be lower at part load than full load.
- Install high turndown burners if frequent on-off cycling is happening.
- Reduce air leaks into the furnace or hot gas leaks from the furnace. These can be from door seals, and openings for conveyors.
- Insulate bare casings.
- Insulate bare piping.
- Preheat combustion air or make-up water from waste heat source or stack gas.
- Preheat the product.
- Automatic boiler isolation valves if piping allows hot water through an "off" standby boiler.

Process Chillers
- Higher efficiency chiller.
- Raise chilled water temperature if possible.
- Lower condenser water temperature for water-cooled chiller.

- Higher efficiency / capacity cooling tower.
- Annual condenser tube cleaning for water-cooled chiller.
- Controls to lock out the chiller below 50 def F.
- Dry coolers or fluid coolers for process chilled water in cool/dry weather instead of chiller.
- Evaporative pre-cooling for air-cooled chillers.

Process Cooling Water
- Use cooling water from a cooling tower or fluid cooler instead of chilled water.

Steam Systems
- Lower steam pressure if possible.
- Implement a steam trap repair program to regularly visit these devices and prevent steam leakage.

Compressed Air Systems
- Locate and repair compressed air leaks.
- Reduce compressed air pressure.
- If source must be maintained more than 20 psi above the end of line point of use, storage or pipe size increase may be appropriate.
- Separate low pressure high volume compressed air uses from higher pressure air, and supply with an air blower instead of an air compressor.
- Pipe compressor air inlet to outdoors, with provisions for low temperature operation.
- Heat reclaim from compressed air heat exchanger (screw), such as ducting to an adjacent area for free heating.
- Where multiple pressure levels are maintained through pressure reduction and a single source, provide separate system for the higher source if it is a lower volume demand than the lower pressure zones.

Humidifiers
- Lower the humidity set point if possible to reduce humidification load.
- Use adiabatic humidifiers (evaporative pads, spray nozzles, atomizers, ultrasonic) if the process combines cooling and humidification.
- Look for inadvertent dehumidification, such as a cooling coil down-

stream of a humidifier.
- Raise chilled water temperature 5 degrees to allow sensible cooling without the side effect of dehumidification.
- Separate humidified and un-humidified areas to limit the humidification load, using a wall and doorway to adjacent areas, to reduce the humidity load.
- Ensure a complete vapor barrier surrounding the space that is humidified, including walls, roof, floor, and penetrations.
- Reduce exhaust if possible, to reduce the humidification load for make-up air.
- Humidity recovery between exhaust and make-up air via 'total heat' wheel with both heat and moisture capture.

Multi-Stage HVAC Air Tempering

Sequential heating, cooling, humidifying, dehumidifying, dew point control, and final tempering operations (e.g. semi-conductor fabrication facilities, printing operations, etc.):
- Closely evaluate each step on a psychrometric chart for opportunities.
- Determine necessary levels of cooling, heating, moisture, and compare to actual operations. Where actual operations go beyond the theoretical levels, ask why and identify the cost implications of doing so. Significant overlap beyond what is theoretically required is very common.
- If the final process is self-heating (semi-conductor fabrication), and upstream supply air is being heated, adjust supply air temperature to reduce load on the final cooling step.
- If dew point control is being used, examine the instrumentation and calibration frequency. Most dew point and relative humidity instruments have a sizeable +/- tolerance, drift over time, and should be calibrated frequently to avoid errors that translate into control overlapping costs. For example, a 5% error in relative humidity at can translate into a 5 deg F error in dew point temperature calculation at 80 deg F and a 10 deg F error 120 deg F which, in turn, drives heating and cooling equipment an additional 5 degrees for naught.
- Where "anticipation" is used for weather changes in large outside air processing operations, evaluate these settings closely since they are often overly conservative at the expense of energy input from unnecessary simultaneous heating and cooling.

Plating Tank Covers
Basis of Savings: Evaporation losses dominate heated tank thermal losses.
- Note: push-pull air curtain systems are for fume control not energy savings. The increased air flow across the tank or vat increases evaporation.
- Even partial covers will reduce evaporation losses.
- **See Appendix "Evaporation Loss from Water in Heated Tanks."**

Process Oven Door Seals
Basis of Savings: Reduced loss from leakage.
- Automated conveyer part baking requires doors be left open. For these, reducing the size of the opening and reducing any ambient air currents that would 'sweep' hot air out of the oven are about all you can do.

Chapter 5

ECM Descriptions

ECM DESCRIPTIONS—ENVELOPE

Envelope Leaks—Infiltration
Basis of savings: Heating and cooling unwanted outside air
- Infiltration from construction cracks usually manifests itself as a comfort complaint, but always increases energy use. In extreme cases it can result in frozen piping. A building pressure test is the perfect solution, but usually not practical. One easy way to check for infiltration leaks is to check return air plenum temperatures during very cold weather. The return plenum is a negatively pressurized area. If there is leakage, it can be found with a hand held infrared thermometer while scanning the perimeter above the ceiling tiles.
- Usually hard to quantify.
- In extreme cases, the leakage around old operable windows can represent more of an energy improvement opportunity than replacing the single pane windows with double pane windows.
- Return air ceiling and shaft plenums are intended to be as air tight as any other duct but are usually far from that. This is especially problematic at building perimeters when the lack of proper construction sealing couples the return air plenum to the building envelope. Since the return air plenum is slightly negative by design, this almost assures infiltration through the envelope. User complaints that point to this are comfort issues at perimeter and cold temperatures above the ceilings at the perimeter. In extreme cases, water pipes can freeze because of this. In humid climates, severe mold damage and sick buildings can be traced to this.

Exterior Color
Basis of savings: Reflecting, instead of absorbing heat.
- **See Chapter 16—Lighting "Reflective Values of Common Colors."**
- If the solar heat gain of the wall or roof can be identified, the difference in reflectivity represents the savings potential.

- For roofs, using a 'pure white' reflectance factor for long-term energy savings is not recommended, since the color will naturally darken with rain and dirt.

Insulation
Basis of savings: Reduced thermal transmission
- A rule of thumb is that 'The first inch of insulation captures 80% of the savings.'
- The biggest opportunities for retrofit are hot or cold surfaces with no insulation at all.
- Increasing insulation thickness during new construction is often cost effective since labor is similar and the material is not expensive.
- Adding insulation to a roof during a roof replacement that will occur anyway is usually cost effective.

Window Upgrade
Basis of savings: Reduced thermal transmission
- Glass is not a very good insulator, but a trapped air space does offer some insulation.
- In all but the most moderate of climates, single pane glazing is a poor choice from an energy perspective. In general, where heating or cooling is needed, single pane units should be avoided and are candidates for replacement if energy use reduction is desired.
- In addition to the glazing, some old window frames are a thermal short circuit. Evidence of this can often be found as water stains on the metal frames from winter condensation. New glazing should always have thermal breaks.
- For replacement of operable windows, poor seals and associated infiltration can pose as much of an energy loss as the windows themselves. New windows should have tight fitting seals.
- Windows are expensive and paybacks are commonly 15-20 years if only energy savings are used to justify the expense. However, window upgrades (the incremental upgrade cost) are easier to justify when windows are being installed anyway.

Window Shading
Basis of savings: Reduced solar load
- Exterior shades are best so that the heat never gets inside.
- Trees, awnings, overhangs, screen covers over skylights.

- Interior shades, if used, should be light colored and highly reflective.
- If existing glazing is clear, coatings or shades can reduce cooling load substantially and can reduce A/C equipment size requirements.
- See Chapter 17—Envelope Information "Glazing Properties" for more information on glass and shading.

Light Harvesting
Basis of savings: Reduced interior lighting
- Portions of the solar load are deliberately allowed into the building to provide day lighting without the use of artificial light. Often this is done near the top of the room enclosure along the wall and reflected off the ceiling. Skylights and clerestories are used to harvest light from the roof to the room below.
- In all cases, the solar gain and added air conditioning load subtract from the lighting savings, although there is usually a net gain.
- For skylights and clerestories, transparent element acts like a hole in the insulation and increases heat loss at that point substantially, especially at night. The envelope heat loss adds heating load and subtracts from the lighting savings. Covers can mitigate this, but are cumbersome and expensive.

ECM DESCRIPTIONS—LIGHTING

Lighting Retrofit: T-12 Magnetic to T-8 Electronic Ballast
Basis of savings: Higher efficacy and reduced ballast loss (lumens per watt)
- Savings of 30% compared to T-12/magnetic ballast are common.
- This is a very common retrofit, since T-12 lights have been the staple fluorescent light installed for a number of years.
- Changing the ballast from "coil and core" (magnetic ballast) to electronic ballast represents the largest part of the savings.
- T-8 bulbs have better quality of light also, in terms of color rendition.
- Proposals to change the ballast but leave the T-12 lights are not recommended, due to compatibility issues. The bulbs need to be changed to T-8 along with the ballast.

Lighting Retrofit: Super T-8 with Low Factor Electronic Ballast
Basis of savings: Higher efficacy
- Special phosphors of these tubes result in greater light output. Without simultaneously reducing the ballast, this system results in the same energy use as a T-8 system with more light.
- But, by carefully pairing these high output bulbs with low "ballast factor" (BF), savings over conventional T-8 lamps with equal light are possible. For example, if the Super T-8 lamps put out 15% more light, and are paired with a ballast selected at BF=0.85 instead of 1.0, there will be approximately equivalent light and 15% less energy consumed, compared to standard T-8 systems.
- Super T-8 bulbs cost more than regular T-8s.

Lighting Retrofit: High Bay HID to High Bay Fluorescent
Basis of savings: Higher efficacy and reduced ballast loss (lumens per watt)
- HID = high intensity discharge
- Savings of 30-50% are common.
- Use caution for fluorescent lights in gymnasium areas since bulbs can shatter.
- High bay fluorescents require reflectors and may be problematic in dirty areas unless the reflectors can be wiped down regularly.
- This is a fixture replacement, usually 1-for-1 to re-use the electrical source.
- Can be T-5HO or, in some cases, T-8.

Lighting Retrofit: Incandescent to Compact Fluorescent (CFL)
Basis of savings: Higher efficacy
- Savings are typically 1/3 to 1/4 the energy for equivalent light, compared to incandescent.
- Installation of self-ballasted CFLs in ceiling 'down lights' is not recommended by manufacturers, since the heat build up will cause premature failure of most standard CFL ballasts. This application could provide reasonable service life if the recessed cans are vented.
- Generally not good for dimming, although advancements continue toward this goal.

Reflectors
Basis of savings: De-lamping with equivalent light
- All light fixtures result in some of the light being "trapped" in the

fixture. Optical reflectors are available as retrofits. Made from highly reflective materials, these push more of the light out of the fixture. In many cases one of the tubes of an existing fixture can be eliminated by virtue of the reflectors. For example, a 3-lamp fixture, retrofitted with reflectors, may provide adequate light with the center tube removed.
- Be certain that the reflector unit is UL listed for a retrofit application.

De-Lamping
Basis of savings: Reduced number of lamps reduces wattage in over-lit areas.
- If it is determined that light levels can be reduced, de-lamping is the simplest of all lighting measures.
- Magnetic ballast may or may not save the amount of energy implied by removing a tube. For example removing 1 of 3 tubes may not reduce energy use by 1/3.
- Most electronic ballasts are capable of operating with one less tube with no harm. If removing more than one tube, consult the ballast manufacturer to be sure. Power reduction with electronic ballast is roughly equivalent to the tubes removed.

ECM DESCRIPTIONS—HVAC

Seal Air Duct Leaks
Basis of savings: Reduced fan energy for a given air flow to the space. For leaks outside the conditioned space, savings are also from reduced unwanted outside air that is heated and cooled.
- Never overlook the obvious.
- In cooling season on a warm day the plenum temperature should be slightly warmer than the space below the ceiling. If it is found to be cooler in the return plenum than in the space on a warm day, then there is almost surely a duct leak.
- Duct sealing is especially important in unconditioned spaces (attics, basements) since air leaked at these points is truly lost. Duct leaks within the insulated envelope are not as critical since the heating or cooling energy is still there providing some benefit during the season.

Correct Control Valves Leaking By Internally
Basis of savings: Eliminating unwanted heating and cooling. In most cases, leaking control valves result in overlapping heating and cooling which doubles the energy waste.
- Testing consists of creating a "full close" command to the device and seeing if the downstream piping or coil returns to ambient or has measurable heating. Sometimes the first row of the coil will be found to be warmer or cooler than the rest of the coil, indicating leaks.
- Electronically actuated valves are especially prone to this due to adjustable travel stops that do not always have good residual close off seating pressure.
- For small piping, 1 inch and less, quarter-turn ball valves can be cost effective and have improved seating quality compared to conventional metal seated globe valves.

Insulate Piping and Valves
Basis of savings: Reduced thermal losses.
- Applies to both hot and chilled water systems, although hot systems represent the higher delta-T and thus the higher heat loss.
- For heating piping, safety is an added justification since many of these represent scald hazards.
- For chilled piping, corrosion protection is an added justification since the condensation that accompanies the cold surface temperature accelerates rust damage.

Lower Chilled Water Condensing Temperature
Basis of savings: Reduced refrigeration cycle "lift"
- 1-1.5% reduction in kW per degree lowered.

Raise Chilled Water Evaporating Temperature
Basis of savings: Reduced refrigeration cycle "lift"
- 1-1.5% reduction in kW per degree raised.

Air-side Economizer
Basis of savings: Avoided refrigeration compressor run time.
- Savings benefit varies by location (available hours of cool air) and by internal load characteristics (hours when cooling is needed while it is coincidentally cool outside).
- The outside air damper and relief dampers should be tightly closed whenever the equipment is off, and relief damper should be closed

ECM Descriptions

tightly in winter when minimum outside air is used, to avoid infiltration in cold weather through these large dampers.
- For economizer mixing boxes that include a relief damper, automatically control it to be tightly closed whenever the outside air damper is at its minimum position, and only begin opening once the outside air damper position moves beyond minimum, lagging behind the outside air damper travel.
- There should be a setting for the economizer operation (usually 55-60 degrees F), below which outside air is deemed sufficient for any needed cooling. Below this point, there should be a positive "cooling lockout" function that prevents compressor operation or forces the chilled water valve fully closed.

Water-side Economizer
Basis of savings: Cooling loads and hours concurrent with low ambient wet bulb temperature can be met with the evaporative effect of the cooling tower directly, without running the chiller.
- The best use for these is if there are steady cooling loads in winter that cannot be served with air-side economizers, often due to the seasonal outside air humidity swings.
- These are inherently not as efficient as an air-side economizer since their use depends on pumps and cooling towers as well as the air handler. However, the chiller does get to shut off so there are substantial savings.

The trouble with most flat plate heat exchanger applications is that
- They are expensive.
- Their capacity is highest when indoor cooling load are usually lowest.
- They are arranged as all-or-nothing, so they are switched off even when they could provide most of the load.
- They share the chilled water system with "non-critical" chilled water loads. Without proper controls (namely, below 55 degrees, no chilled water to these units), the system reaches the cut off point sooner than it would need to.
- Note that there are some piping and pumping arrangements that will allow the flat plate system to run concurrently with the chiller, normally to pre-cool the chilled water return. This is a complex design solution but may be effective if the load and wet bulb temperature

CONVENTIONAL AND EXTENDED FLAT PLATE OPERATION					
	All Possible Flat Flat Plate Hours	approx. pct full load capacity	6a-6p annual hours	24-7 annual hours	Outside Air Conditions
EXTENDED	2.5 deg pre-cool	25%	180	480	OA 40-42.5 wb
EXTENDED	5 deg pre-cool	50%	240	630	OA 37.5-40 wb
EXTENDED	7.5 deg pre-cool	75%	60	150	OA 35-37.5 wb
CONVENTIONAL	chillers OFF	100%	1260	2820	OA <35 wb and <55 db
	increased flat plate hours		38%	45%	
Based on Wet Bulb Temperature Profile, Colorado Springs.					

Figure 5-1. Water-side Economizer Flat Plate Hours for One City

profiles show potential by stretching out the flat plate cutoff point. Sharing a single cooling tower, this requires some form of compensation for the chiller, since it cannot run on excessively cold condenser water. The accommodation may be throttling the condenser water flow to maintain head pressure on the chiller, or with recirculating/blending pump at the chiller condenser water inlet to temper the cold water. Where additional pumps are used to accommodate dual operation, the benefit of the partial flat plate operation must be compared with any additional pumping energy expense.

Angled Filters Instead of Flat Filters
Basis of savings: Reduced average pressure drop reduces air horsepower in VAV systems and in CAV systems if re-balanced.
• Can be combined with early change out for increased benefit.

Bag Filters instead of Cartridge Filters
Basis of savings: Reduced average pressure drop reduces air horsepower in VAV systems and in CAV systems if re-balanced.
• Can be combined with early change out for increased benefit.
• Filter manufacturers caution on "dirt release" from bag filters when stopped and started regularly, therefore this measure is best suited for fans with extended run times.

Multi-zone Conversion to VAV
Basis of savings: Reduced overlapping heating and cooling and reduced energy transport work since air flow is proportional to load instead of constant volume.
• Hot deck is blanked off and zone mixing dampers sealed in the "cold

deck" position. VAV boxes added for each zone, usually near the air handler and re-use the zone ductwork.
- A sketch of this modification is in the **Appendix "HVAC Retrofits of the Three Worst Systems."**

Multi-zone: VAV Conversion Using Existing Zone Dampers
Basis of Savings: Reduced heating-cooling overlap and reduced fan energy.
- Main fan is controlled on static pressure in the hot/cold deck plenum box.
- Existing zone dampers act as pressure dependent VAV dampers. Uses existing ducts for economy.
- Linkage between hot and cold dampers must be split and actuators provided for independent hot and cold damper control.
- Unless the zone mixing dampers are in good condition and tight sealing, it may be more cost effective to convert to VAV.
- Care must be used to provide minimums and ventilation air.

Multi-zone Conversion to Texas Multi-zone
Basis of savings: Reduced overlapping heating and cooling penalty due to the introduction of the "bypass" air path.
- This conversion makes the hot deck of a multi-zone unit into a bypass deck.
- Reheat coils are added for each zone downstream of the mixing dampers.
- The air stream delivered to the zones can be either cooled, warmed, or bypass (re-circulated), compared to either warmed or cooled.
- System remains constant volume.

Dual Duct Conversion to Separate Hot Deck and Cold Deck Fans
Basis of savings: Reduced heating burden in the hot deck.
- Independent control of hot duct and cold duct in the zone mixing boxes reduces the inherent overlap in these systems.
- Independent hot and cold duct systems via separate fans allow seasonal optimization for hot and cold decks. For example, mixing outside air to reduce cooling cost in winter adds to heating cost since both ducts share the same mixed air stream.
- Introduce the ventilation air into only one of the ducts—normally the cold duct.
- The hot deck simply re-circulates and does not see the cold ventilation air as a load.

- Additional savings are possible by converting from constant volume to variable volume in both ducts.
- A sketch of this modification is in the **Appendix "HVAC Retrofits of the Three Worst Systems."**

Spot Cooling
Basis of savings: Heat only the worker or process, not the whole factory.
- Concepts of "air changes" and "cfm per SF" do not apply.
- Duct size, cfm, and heat/cool energy delivered is much smaller.
- Can be very beneficial for un-insulated buildings, high bay buildings, and factories with a lot of stationary heat producing equipment (ovens, kilns, laundry, etc).
- Special design considerations for this system to be effective.

Evaporative Pre-cooling for Air-cooled Condensers
Basis of savings: Reduction of dry bulb temperature from the evaporative cooling process lowers the condensing temperature and reduces refrigeration cycle lift. The air-cooled equipment "thinks" it is cooler outside and behaves accordingly.
- Approximately 1-1.5% kW reduction per degree reduction.
- Economic break even point starts around 50 tons.
- Water and waste costs compete with energy savings. Drain, fill, and freezing are design considerations.

This diagram shows the effect for air cooled equipment in a climate with design temperatures of 95 deg F dry bulb and 58 deg F wet bulb. Similar benefits are possible in different climates provided the wet bulb depression value is similar.

Adiabatic Humidification
Basis of savings: Evaporation without energy input to cause boiling.
- Includes evaporative pads, spray nozzles, atomization, and ultrasonic methods. Good results if combined with a process that simultaneously needs cooling. No energy advantage if the cooling effect is not beneficial and must be counteracted with new-energy heating.
- Psychrometric process follows the constant enthalpy line and increases moisture content as temperature drops.
- Compressed air and ultrasonic technologies each use approximately $1/10^{th}$ of the energy compared to boiling the same amount of water.

Note: all of the adiabatic evaporative methods require air at a reasonable temperature to humidify and generally will not work well in

ECM Descriptions

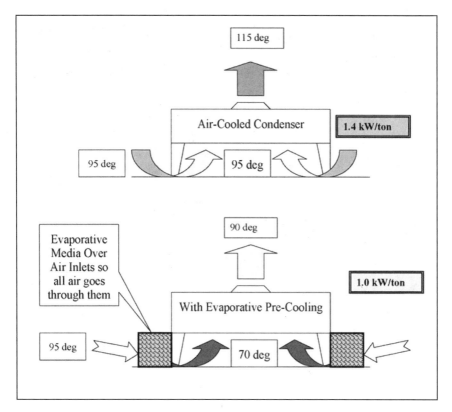

Figure 5-2. Evaporative Pre-cooling

air temperatures lower than 50 deg F, therefore application is best suited to the return air stream or other tempered air stream and not in the outside air stream in cold weather.

Adjacent Air-cooled Equipment Too Closely Spaced
Basis of Savings: Correcting re-circulated discharge air condition lowers inlet air conditions so the air-cooled equipment energy use is lowered as if it were a cooler day.
- Warm air re-entrainment can occur from improper equipment spacing. Elevated intake cooling air directly raises refrigeration head pressure and compressor power by 1-1.5% per degree.
- A good rule of thumb for proper spacing is the air inlet or vertical finned coil height projected horizontally.

Figure 5-3. Improper Air-Cooled Condenser Spacing
This installation was measured with 5 degrees of elevated inlet air temperature and an estimated power increase of 5-7% as a result.

Constant Speed to Variable Speed Pumping Conversion
Basis of Savings: Pump energy use profile matching the heating/cooling load profile will use less energy than "constant volume" constant energy pumping.
- The key is to provide the fluid flow upon demand, but not all the time. For comfort systems, the system will be controlled to track the load profile. Applies to HVAC applications including chilled water pumping, condenser water pumping, and heating water pumping. Requires a variable speed controller for the pump and a load following signal.
- For example, if chiller load (tons) is known, the adjusted flow rate and speed for pumps can be derived automatically to proportionally follow the load change. The most substantial changes will occur at hours when load is between 50% and 100% load (flow between 50-100%).
- Additional flow savings can be obtained by increasing system delta T (dT), which allows the thermal energy transport to occur with less

mass flow and pump work.
- The savings of pump energy should be weighed against any change in compressor efficiency to be sure there is a net gain, especially from condenser water.
- Chilled water flows below 2 feet per second often encroach on laminar flow and can result in compressor energy penalties that negate pump savings. The prudent approach is to verify for the specific chiller the compressor kW/ton at the proposed changed flows and temperatures to assure there is a system benefit and not just a pump benefit.

Constant Volume to Variable Air Volume Conversion
Basis of Savings: Fan energy use profile matching the heating/cooling load profile will use less energy than "constant volume" constant energy air moving.
- Applies to any air moving system, including exhaust, supply, and make-up. Applies equally to process air movement as comfort systems. The key is to provide the air flow movement upon demand, but not all the time.
- The energy savings is from having fan energy track the load profile. Applies to comfort air conditioning and heating or any other air moving task with a variable load that can be served by varying the volume of air. Requires a variable speed controller for the fan and a load following signal, commonly a downstream pressure sufficiently high to allow VAV boxes to operate.
- For a given heat load (heating or cooling) and differential temperature, the required air flow can be easily calculated and is seen to directly follow the changing load. The most substantial changes will occur at hours when load is between 50% and 100% load (flow between 50-100%)
- Additional flow savings can be obtained by increasing system differential temperature (delta T), which allows the thermal energy transport to occur with less mass flow and fan work.

Constant Volume Terminal Reheat to VAV Reheat Conversion
Basis of Savings: Reduced overlapping heating/cooling and reduced fan horsepower.
- Each zone reheat coil is replaced with a VAV box. Most will require a heating coil unless serving only an interior area.
- Careful evaluation of the upstream duct system is required to be cer-

tain it will be suitable for the higher pressures involved.
- Heat is available only after the air flow has been reduced to "minimum cooling flow," therefore less reheating of supply air occurs.
- A sketch of this modification is in the **Appendix "HVAC Retrofits of the Three Worst Systems."**

Testing Adjusting and Balancing (TAB)
Basis of Savings: Reduce system pressure by excessive damper throttling, thereby reducing transmission energy requirements.
- This measure requires knowledge of existing positions of dampers or valves. In some cases, if operations staff or maintenance procedures over the years have spoiled the original balancing effort, the balancing status maybe a large unknown quantity and this measure may be as much about restoring confidence and performance as anything else. But, while it is being balanced, do so in a way that encourages low fan/pump energy input. Also, be sure to have locking provisions and permanent marks for the balancing devices, to help sustain the work.
- Impeller trimming for over-sized pumps can be a big energy saver, equal to the amount of energy dissipated at the balancing valve.

Proportional Balancing Method
Described for air, but is applicable to both water and air balancing.
- The system is first measured with all dampers fully open and the fan at full output.
- With the initial iteration, each outlet is measured and given a percentage of the design intended flow. Some may be above and some may be below the intended flow rates.
- The outlets are numbered in order of increasing percentage of design airflow. The outlet with the lowest percent remains open and is not adjusted. Its percentage is designated as B and is the basis for other branch adjustments.
- The flow rate on the branch with the next lowest percentage is adjusted so that it has the same percentage as B. All other branches are adjusted to this same value of B.
- When that initial step is complete, then adjust the main fan capacity (by dampers, sheaves, motor speed, motor change, etc.) to achieve full design capacity.
- Then return to the individual outlets and spot check at least 20% of them to assure that they are within the stated tolerance. If they are

not, then repeat this process iteratively until they are. At the conclusion, there is still at least one damper that is fully open, to minimize overall system loss.

Remove Inlet Vanes or Discharge Dampers
Basis of Savings: Reduced System Pressure and Fan Hp.
After converting to VFDs, these devices create unwanted air pressure drop even in the wide open position.
The wide-open loss depends on the free area (FA) of the damper and the air velocity. Inlet vanes and control dampers are typically around 80% FA.

Inlet Vane and Outlet Damper Pressure Drop Loss
Source: American Warming & Ventilating
Based on the following relationship and free area factors.

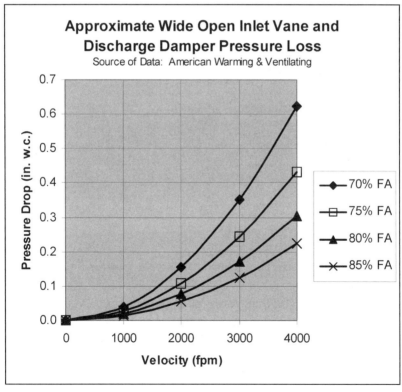

Figure 5-4. Inlet Vane and Outlet Damper Pressure Drop Loss

$$(\text{FA Factor})(\text{Velocity}/4005)^2$$

FA Factors
70% - (0.624)
75% - (0.434)
80% - (0.306)
85% - (0.224)

Lower VAV Duct Static Pressure Control Setting
Basis of Savings: Power is reduced exponentially with a reduction in pressure.
The relationship $HP2 = HP1 \times (SP2/SP1)^{y1.5}$ is a derivation of the fan laws that applies to VAV systems without the constraint of a maintained downstream duct pressure. However, most commercial VAV systems use a maintained downstream duct pressure and so this relationship must be de-rated. This de-rate is required since the pressure reduction is only occurring in the discharge duct section—losses in the balance of the system, including casing and coil losses, are not affected. So, only external losses contribute to this and of those, only the discharge section.

The exact value of the exponent depends on the proportions of internal to external duct resistance and the fraction of the external duct resistance impacted by the set point reduction.

Fraction of total air system resistance affected by set point reduction	Exponent
100%	1.5
75%	1.1
67%	**1.0**
50%	0.8
25%	0.4

Figure 5-5. Fan Law Exponents for VAV Static Pressure Reduction

A short cut method that works for most standard VAV air handlers is to drop the exponent, such that the savings from reduced pressure varies proportionally with the pressure reduction. This assumes 2/3 of the fan energy is spent on the supply duct—which, by design is high velocity and restrictive and the most energy intensive portion of the duct system.

$$HP2 = HP1 \times (SP2/SP1)$$

Reduce Resistance in Distribution Ducts and Pipes

Basis of Savings: Power is reduced with a reduction in resistance.
This model considers removing an obstruction or otherwise 'lowering the bar' of the whole system for pressure requirement and applies to both constant and variable flow.

Before considering changes to equipment and controls, consider the source and the distribution system. Begin by using less if possible: less air, less water. Do you need all that air or water circulating?

Possible ways to reduce system resistance:

Water
- Remove balancing valves after conversion to VFD
- Re-balance so at least one balancing valve is fully open
- Clean strainers earlier
- Replace restrictive main line strainers
- Replace restrictive main line flow meters (orifice plates)

Air
- Remove old inlet vanes or discharge dampers after conversion to VFD
- Re-balance so that at least one balancing damper is fully open
- Clean coils
- Change filters earlier
- Extended surface or angled filters
- Eliminate unnecessary duct appurtenances
- Smooth out 'bad' duct fittings

Power Reduction from Reduced Duct and Pipe Friction

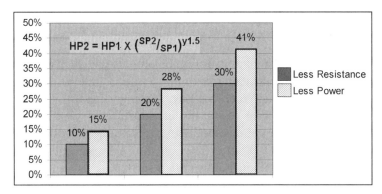

Figure 5-6. Fan Law Exponents for VAV Static Pressure Reduction

Correcting Bad Duct Fittings, Entrance Losses, Exit Losses
Basis of Savings: Reducing system pressure reduces fan horsepower requirements.
The basic relationship for air horsepower is

$$HP(air) = CFM * TSP/6356$$

Where CFM is cubic feet per minute air flow and TSP is the total static pressure, in. w.c.
Motor Hp is higher than air horsepower, because losses from fan efficiency, fan drive, and motor efficiency must be incorporated.

See **Chapter 11—Mechanical Systems** "Fan/Pump Motor Work Equation" and "Fan and Pump Efficiencies, and Belt Drive Efficiencies"

See **Chapter 12—Motors and Electrical Information** "Motor Efficiencies"

For a given air flow, reducing TSP directly reduces horsepower. In addition to the friction losses, coil and filter pressure drops etc., part of the work of the fan is fittings and entrance/exit losses. When one of these is especially bad, the pressure drop can be unusually high and an opportunity for improvement.

This book will not fully cover duct fitting losses. There are a number of excellent texts on this, namely *ASHRAE Fundamentals and SMACNA Duct System Design Manual.*

Some Rules of Thumb:
- Transitions should be no more than 30 degrees, and 15 degrees per side is better.
- Abrupt (blunt) changes in duct geometry are usually bad, unless velocity is very low (<1000 fpm)
- Square elbows should always have turning vanes
- Velocity is usually highest at the fan outlet
- Fan discharge conditions are high loss areas unless they have smooth transitions
- Losses increase exponentially with velocity
- Many bad duct fittings are there because of lack of room and don't have good solutions

- HVAC air velocity over 2500 fpm requires special care in design to avoid high losses

Losses are normally calculated in terms of "number of velocity heads" (Hv), where tables provide the "C" factor based on testing.

$$Hv\ (air) = (V/4005)^2$$

where Hv is in. w.c., and V is velocity in feet per minute

Example:
A poor duct fitting has a C-factor of 0.9 and can be replaced with one having a C-factor of 0.20. What are the savings in hp if the air flow is 20,000 cfm and the duct velocity is 2000 fpm at that point?

Ans:
Hv = $(2000/4005)^2$ = 0.25 in. w.c.
Savings is the differences of "C," which is 0.9-0.20 = 0.7
Reduced pressure is 0.7 velocity heads, or 0.7 * 0.25 = 0.175 in. w.c.
HP = (20,000 * 0.175)/6356 = 0.55 hp

Often bad duct fittings are accompanied by noise, either from the fitting or from the fan laboring to move air through the turbulence.
See **Appendix "Duct Fitting Loss Coefficients"** for a list of some bad duct fittings and their "C" factors.

Distributed Heating instead of District Heating
Basis of Savings: Reduced thermal losses at part load.
There are some good reasons to use district heating.
- Centralized maintenance
- Aesthetics
- Possible source of free heat source to utilize

However, energy of this system is usually higher than for a distributed heating system, where heating equipment is located closer to the point of use. The main reason for this is the thermal line losses. While these may be kept to a tolerable level at design loads, unless the heating fluid temperature is reset in mild weather, these losses are constant and form a larger and larger fraction of the total energy use at part load.

Figures 5-7 and 5-8 show overall system loss at part load, given full load thermal losses. These losses (2%, 5%, 10%, 20%) represent total distribution thermal losses (piping losses).

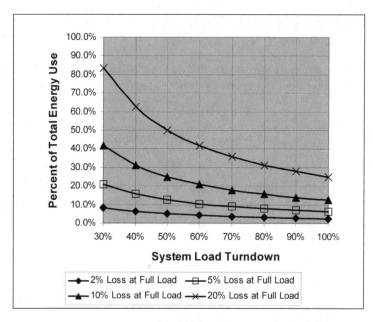

Figure 5-7. Thermal Losses at Part Load (80% firing efficiency)

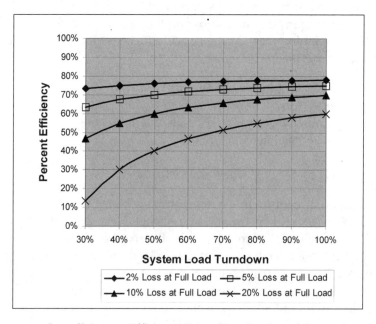

Figure 5-8. Overall System Efficiency Loss at Part Load (80% firing efficiency)

ECM DESCRIPTIONS—SWIMMING POOLS

Pool Covers
Basis of savings: Isolates the water surface from air, preventing evaporation.

Reduce Pool Evaporation
Basis of savings: Water lost through evaporation absorbs heat, 1000 Btu per pound, cooling the remaining water. Additionally, the water lost is replaced and must be heated to pool temperature. Reducing evaporation reduces these losses directly.

Pool energy use is mostly due to water heating, and heating burden (other than initial heating) is largely due to evaporation, so controlling evaporation and other water losses is an important consideration in an energy program. Evaporation is strongly affected by two things: *wind speed* at the surface and the *differential vapor pressure* between the water and air, so energy use from pool evaporation can be reduced by:

- Lowering the pool temperature
- Raising the surrounding air temperature
- Raising the surrounding air humidity
- Lowering the wind speed at the air-water surface interface

Note: HVAC systems that exchange air to dehumidify should do so based on relative humidity controls, and excessive drying by using outside air will increase evaporation.

Recommended Pool Water and Air Temperatures
Surrounding air temperature should generally be higher than and within 2 degrees of the water temperature to reduce evaporation, although air temperatures over 85 def F are not recommended for comfort reasons. Higher humidity lowers evaporation rates, although humidity over 60% is not recommended to reduce risk of biological growth.

Figure 5-9. Recommended Pool Air and Water Temperature

Type of Pool	Air Temperature, deg F	Water Temperature, deg F	Relative Humidity, pct rH
Recreational	75-85	75-85	50-60
Therapeutic	80-85	85-95	50-60
Competition	78-85	76-82	50-60
Diving	80-85	80-90	50-60
Whirlpool/Spa	80-85	97-104	50-60

Source: ASHRAE Applications Handbook, 1999, © American Society of Heating, Refrigerating and Air-Conditioning Engineers, Inc., www.ashrae.org.

Pool Evaporation Facts

Actual values will vary depending upon individual conditions, but the basic relationships can be illustrated as follows:

- At 50% rH, increasing wind speed over the water increases evaporation rate 30% for each 1-mph of wind speed increase.
- At 50% rH, reducing surrounding air temperature increases evaporation by 4% for each 1-degree lowered.
- At 60% rH, reducing air relative humidity over the water increases evaporation 2-3% for each 1-pct rH lowered.
- Mechanically heating the pool air over the water is 15% more efficient than using the pool to heat the surrounding air, due to reduced evaporation.

Pool Evaporation Formula

Simplified Pool Evaporation Formula
ASHRAE 1999 Applications Handbook

$wp = 0.1A * (pw-pa) * Fa$

wp = evaporation of water, lb/hr
A = area of pool surface, ft^2
pa = saturation pressure at room air dew point, in. Hg
pw = saturation vapor pressure taken at surface water temperature, in. Hg
Fa = Activity Factor:

Residential pool:	0.5
Condominium	0.65
Therapy	0.65
Hotel	0.8
Public, schools	1.0
Whirlpools, spas	1.0
Wave pools, water slides	1.5

Figure 5-10. Simplified Pool Evaporation Formula

ECM DESCRIPTIONS—HEAT RECOVERY

General Criteria

The term "heat recovery" is the commonly used term for a variety of systems that recovery a portion energy from a waste stream and put it to use, displacing new energy consumption that would otherwise occur. "Energy recovery" would be a more descriptive term since the concept applies to more than just heat. The most common application in industry is taking waste heat and using it to heat or pre-heat another mass that needs heating anyway, thereby saving fuel. The same principles apply for air or water flows needing to be cooled anyway—cooling or pre-cooling from waste streams that are cooler. Humidity recovery is also viable for large air streams by pre-humidifying from a more moist exhaust air stream, or de-humidifying make-up air with a drier exhaust air stream. In all cases, heat recovery serves to reduce new energy input by re-using waste products. While the goal is admirable, there are a number of practical barriers.

Heat Recovery Viability Test

For heat recovery to viable, it must meet these requirements:

1. *Same Time.* Generally, waste heat and the need for heat must occur concurrently. Thermal storage could assist this, although this is rare. To illustrate, drain waste pre-heating domestic hot water makes sense for a shower, but not for a bath.

2. *Right Proportions.* The only amount of recovered heat that counts is where the available heat and needed heat are equal. When there is an abundance of either waste heat or needed heat, it is the smaller of the two that determines the actual energy recovery effect. To completely eliminate the new-energy heating and supplant it entirely with waste heat, the waste heat quantity must exceed the need for heat recovered heat. Best economics occur when the two are in similar proportion.

3. *Right Temperatures.* Waste heat must be at a sufficiently higher temperature than the recovered heat sink to allow heat transfer. The greater the differential temperature available, the more economical the apparatus due to the higher "approach" temperature there is to work with.

4. *Enough Recovered Heat to Make it Pay.* There must be sufficient heat transfer throughout the year to provide economic justification for implementation. One rule of thumb is 5000 CFM of outside air intake with equal exhaust nearby. Single pass systems (outside air make-up, raw water heating) are good candidates since they are energy intensive. Facilities with a large number of operating hours per hear, especially continuous operations, provide quicker returns than others.

Heat Recovery Application Notes
- New construction can benefit the most from heat recovery proposals since the cost of the recovery equipment is subsidized by the reduced size and cost of the primary heating equipment that would otherwise be provided. The down side to this is that the facility now becomes dependent upon the heat recovery system, and so its proper sizing, application and maintenance for sustained operation become more critical.
- For retrofits the heating equipment is backup and the recovery benefit simply reduces its load. Savings in energy alone must pay for the recovery equipment and so returns are longer.
- Recovered heat energy is penalized from additional air horsepower from pressure drop through coils, pumping energy, standby losses, and from capital investment requirements.
- When either the waste heat or recovered heat temperatures vary over time, there can be times when heat transfer is marginal, or when heat recovery is detrimental. Whenever the differential temperature is near zero, the actual heat transfer will be very low and the energy used by active system components (fans, pumps) may not be justified. Controls monitoring the differential temperatures should be used to determine when to shut down the system.
- For *refrigeration* system heat recovery, a good rule of thumb is 4000 Btuh per ton of refrigeration capacity available for hot water heat recovery.

Exhaust–to–Make up Air
Basis of savings: Reduced heating and cooling energy for make-up air, compared to 'raw' outside air.
- Examples:
  - Gas clothes dryer vent used to pre-heat incoming combustion air.

ECM Descriptions

- Paint booth exhaust pre-heating make-up air.
- Building exhaust pre-heating make-up air.

Rejected Heat-to-make-up Water
Basis of savings: Reduced heating energy for make-up water, compared to 'raw' make-up water.
- Examples:
- Boiler economizer, pre-heating make-up water.
- Refrigeration hot gas (before going to the condenser), pre-heating make-up water.

Rejected Heat-to-space Heat
Basis of savings: Reduced heating demand on the space heating system.
- Examples:
- Waste heat off an air compressor that can warm a section of a factory.
- Refrigeration system rejected heat used as heat in an air handler or as reheat in a dehumidification cycle.

Wastewater-to-make-up Water
Basis of savings: Reduced heating energy for make-up water, compared to 'raw' make-up water.
- Example:
- Commercial laundry waste water heat exchange to pre-heat wash water.
- Injection mold cooling jacket water to warm process water or boiler feed water.

Combined Heat and Cool: The Water-to-Water Heat Pump
Basis of savings: Compound savings: cooling energy saved and heating energy saved. Usually expressed as cheap heating and free cooling.
- Waste refrigeration heat can be used as a primary heat source, provided there is a concurrent need for both heating and cooling. The chiller pre-cools the chilled water return, shedding load on the main chillers, and the condenser becomes a water heater.
- Special equipment is needed to achieve refrigerant hot gas temperatures sufficient to make space heating water temperatures of 140 def F or higher. Coefficient of Performance values (COPs) go down at these temperatures but are still around COP=2.0 which is not great for a cooling machine but is very respectable for a heating machine.

Recovery of Humidified or De-humidified Air

Basis of savings: Where humidified or dehumidified exhaust occurs, the energy normally used for humidifying or dehumidifying the make-up air can be recovered.

- Special heat wheels with desiccant or other moisture holding material are used.

Examples:
- Building exhaust pre-heats and pre-humidifies outside air intake in winter.
- Building exhaust pre-cools and pre-dehumidifies outside air intake in summer.

Type	Efficiency (note 1)
Heat Wheel	60-80%
Heat Pipe	60-70%
Plate-Box	60-80%
Run-Around Coil	40-60%

Figure 5-11. Air-side Heat Recovery Equipment Efficiency Guidelines

Note 1: thermal heat recovery potential. Does not include parasitic losses of fan or pumps, or added resistance of the recovery equipment in the fluid path. There are a number of variables that determine application efficiency such as relative temperatures and proportions of flow rates. Values vary by manufacturer—for example some low end commercial heat recovery wheels may be less than 50% efficient.

Double Use of Process Air and Water in Heat Recovery

Basis of savings: Both the energy and fluid itself are re-used directly, without heat exchange apparatus. The second point of use is seen as 'free'. Note that consideration of contaminants as well as implications of failure and shutdown of individual equipment is required.

- General exhaust from a theater used as make-up and cooling for a projector
- General exhaust from a building used as make-up for toilet exhaust or kitchen exhaust

- Steam condensate not being returned added directly to wash water
- Waste heat exhaust from an air compressor intercooler used for boiler make-up air in winter
- Single pass refrigeration cooler water used for cooling tower make-up

ECM DESCRIPTIONS—THERMAL STORAGE (TES)

Basis of savings: Savings for all thermal energy storage cooling systems is the ability, via storage, to use energy during off peak times when it costs less.

TES Pros and Cons
- Full storage systems are capable of keeping the chiller off the entire on-peak time. The storage systems cost more since the number of ton-hours is higher. However, conversions of existing conventional chiller plants to TES may be sufficiently sized and are candidates for full storage if the refrigeration equipment is in good condition with life remaining in the equipment.
- Partial storage TES systems serve to defray part of the on-peak demand and flatten the electrical load profile. They run concurrently with the refrigeration system during the day and run at night to re-charge the storage. This means the refrigeration equipment runs almost continuously. Still, these systems, especially on new construction, are less expensive to install.
- Flexibility is a key detriment to most of these systems. Even a properly sized system can be rendered obsolete if the rates change, and the chances of rates changing during a 20 year equipment life cycle are very good.
- Cool storage and warm storage systems are all plagued with stratification losses. Various attempts have been made to deal with it, and it remains an engineering challenge.
- Systems with cyclic freeze-thaw are often plagued with expansion damage. Design must include ample provision to accommodate these forces.
- For warm or cool storage, the container volume is an order of magnitude larger with correspondingly greater surface area (compared

to ice), and stand-by losses become increasingly important. Standby losses are probably on the order of 20% for warm or cool storage.
- Minimum of 25% energy penalty in ice making mode, even with lower condensing temperatures at night.
- TES normally only makes economic sense when there is a large rate incentive for off-peak use, and never makes sense if energy conservation is important. If the rates are there, these systems can save utility bill money, but almost always use more energy, and almost always cost more to install.
- Significant off-peak utility rate discounts for energy and demand are usually required to make such systems attractive. However, other considerations may make thermal storage a good choice, such as ride-through back up for critical cooling applications, allowing hours of cooling in the event of a power loss while using generator power for circulating pumps only.

Rules of Thumb for TES Systems
- Cool storage: 100 gal/ton-hr (15 deg dT), 150 gal/ton-hr (10 deg dT)
- Encapsulated ice: 17-22 gal/ton-hr
- Ice on coil: 18-26 gal/ton-hr
- Installed TES system (ice), all types: approx $100 per ton-hour. Source: Cryogel, 2007.

Conditions Favoring Thermal Energy Storage
- Average cooling loads are much less than the peak cooling load.
- Large differential between on-peak and off-peak energy and demand charges.
- Low off-peak demand charges.
- High number of seasonal cooling (or heating) hours and ton-hours load.
- Utility incentives to defray first cost.
- Available space for storage containers.
- Higher-than-average operational staff technical expertise.

Cool Storage
- Chilled water is created during off peak times when power costs are less. The chilled water is stored in a large tank and used during the day allowing the chillers to be turned off.

ECM Descriptions

TES option comparison	Conventioal mechanical refrigeration	FULL STORAGE chilled brine ice storage	PARTIAL STORAGE chilled brine ice storage	FULL STORAGE chilled water cool storage	FULL storage evaporative cooling with cool storage
Max cooling load, tons	500	500	500	500	500
Annual load, ton-hrs/yr	720,000	720,000	720,000	720,000	720,000
Storage stand-by loss %	0%	5%	5%	20%	20%
% system storage	0%	100%	50%	100%	100%
Storage stand by losses	0	36,000	18,000	144,000	144,000
Total cooling load, ton-hours	720,000	756,000	738,000	864,000	864,000
Tons capacity	500	350	175	350	350
hours of storage	10	10	10	10	10
Design daily storage load, ton-hrs (tons*hrs*storage factor, incl storage loss)	0	5,250	2,625	6,000	6,000
Storage tank size, gal	N/A	95,000	47,500	600,000	600,000
Chjiller kW/ton	0.60	0.75	0.75	0.60	0.00
Auxiliary kW/ton	0.20	0.30	0.30	0.30	0.30
Total kW/ton	0.80	1.05	1.05	0.90	0.30
Total kW demand (kW/ton * installed tons)	400	368	184	315	105
Annual energy use, kWh (ton-hrs * kW/ton)	576,000	793,800	774,900	777,600	259,200
kWh on-peak	200,000	0	150,000	0	0
kWh off-peak	376,000	793,800	624,900	777,600	259,200
kW demand on-peak	400	0	184	0	0
kW demand off-peak	400	368	184	315	105
kW demand off peak above on peak	0	368	0	315	105
$/kWh on peak utility cost	0.06	0.06	0.06	0.06	0.06
$/kWh on peak utility cost	0.03	0.03	0.03	0.03	0.03
$/kW-yr on-peak utility cost	120	120	120	120	120
$/kW-yr off-peak utility cost	80	20	80	20	20
$ for kWh on-peak	12,000	0	9,000	0	0
$ for kWh off-peak	11,280	23,814	18,747	23,328	7,776
$ for demand on-peak	48,000	0	22,050	0	0
$ for demand off-peak	0	7,350	0	6,300	2,100
$ Total elec cost	71,280	31,164	49,797	29,628	9,876
M$ Total 20-yr elec cost	1.43	0.62	1.00	0.59	0.20
$ Maintenance cost per year	50,000	50,000	40,000	30,000	20,000
M$ 20-yr maintenance cost	1.00	1.00	0.80	0.60	0.40
M$ Installed cost premium	0.0	0.4	0.0	0.6	0.1
M$ Total 20-yr cost	2.43	2.02	1.80	1.79	0.70

Figure 5-12. Sample TES Cost Comparison

Notes:
1. Parameters will vary by locale. This is intended to show how the factors to compare.
2. No first cost incentives considered.
3. For partial storage, any number of fractions of storage are possible. 50-50 was used for this example.
4. Cost of water considered equal for each option and not shown.
5. For full storage systems, a special thermal storage electric rate is assumed.

Ice Storage
- Ice is created with low temperature brine during off peak times when power costs are less. The chilled water is stored in a large tank and used during the day allowing the chillers to be turned off.

Phase Change Material (PCM) Storage
- Same as ice storage, except PCMs can be selected for phase change at temperatures closer to utilization temperature (45 def F) instead of ice (32 def F). This technology has the potential to leverage the compactness advantage of ice storage equipment without the 25% inherent energy penalty.
- The barrier is that PCMs are costly while water is not.

Cool Storage—Evaporative Cooling
- In dry climates, it is possible to use evaporative cooling at night in conjunction with cool storage, to reduce or eliminate mechanical refrigeration. This is a very good way to save energy in refrigeration—by turning it off completely—however pumping costs will be higher and will offset some of the savings. Cooling season wet bulb temperatures in the mid-40s and low 30s are needed to drive this. In dry climates, this variation has strong promise.

ECM DESCRIPTIONS—ELECTRICAL

Power Factor Correction
Basis of savings: There are two.
(1) Electric utility charges for power factor, and
(2) Decreased I^2R copper loss within the facility distribution wiring, from the wires carrying the excess magnetizing currents.
- Utility fees alone can often justify power factor correction projects.
- Example: For some utilities, the power factor charge is a 1-for-1 increase in demand charge. If the power factor is habitually 80 percent and the utility charges for anything lower than 95 percent, then the cost of poor power factor in this case would be (95-80) = 15 percent increased demand charges.
- Note: Simple capacitor installation may bring good results for facili-

ties with mostly motor loads, welding, etc. But in facilities with high levels of harmonics should be studied very carefully and special and costly power factor correction equipment may be required.
- See **Chapter 12—Motors and Electrical Information "Power Factor Correction Capacity Quick Reference Chart."**

Load Balancing
Basis of savings: Motor performance and efficiency presumes equal voltage on each phase. If the voltage and current are different, then one phase will pull harder than the rest, and the motor windings fight, with ensuing energy loss and motor heating.
- A 2 percent voltage imbalance on a polyphase motor can reduce efficiency by 5 percent. For example, a motor with an 85 pct eff. nameplate could be 0.85 * 0.95 = 80.75 percent efficient.
- See **Chapter 12—Motors and Electrical Information "Voltage Imbalance Effect on Motor Efficiency"** for motor efficiency losses at other degrees of imbalance.

ECM DESCRIPTIONS—OTHER

Lower Compressed Air Pressure
Basis of savings: Reduces compression ratio and required compressor work.
- Approximate savings are 1% power for each 2 psi lowered.

Lower Compressed Air Inlet Temperature
Basis of savings: Increases inlet air density, increasing volumetric efficiency and required compressor work.
- Approximate savings are 1.9% power for each 10 deg F lowered.

Lower Steam Pressure
Basis of savings: Several Factors:
- Lower steam pressure means lower steam temperature, and heat equal heat transfer occurs at reduced combustion temperatures. Thus, reduced pressure reduces stack temperature and casing radiation losses.
- Reduces feed water pressure and pumping requirements

- Reduces losses from leaks
- Reduces condensate temperature and attendant losses
- See also **Chapter 13—Combustion Equipment and Systems, "Estimated Savings of Steam System Improvements."**

Lower Boiler Excess Air
Basis of savings: Reduces the cooling effect of the extra air, increasing combustion temperature and heat transfer. Excess air 'sweeps' heat out of the boiler with no benefit.
- See **Chapter 13—Combustion Equipment and Systems, "Estimated Savings from Reducing Excess Air."**
- Approximate 1% efficiency gain for each 15% reduction in excess air.

Pre-heat Combustion Air
Basis of savings: Increases combustion temperature and heat transfer.
- Approximate savings are 1% efficiency increase for each 40 deg F pre-heat.

Pre-heat Feed Water
Basis of savings: Reduces the cooling effect of the feed water on the boiler water, reducing heating load.
- Approximate savings are 1% efficiency increase for each 10 deg F of pre-heat.

Chapter 6

Utility Rate Components

Note: For heating fuel, only natural gas is discussed here.

ELECTRIC

Common utility rates are summarized here. Infrastructure fees, administrative fees, special customer rates, extremely large customer rates, cogeneration, net metering, wheeling, and de-regulated direct purchase rates and surcharges to fund conservation programs are not included.

Energy
Typical Units: $ per kWh or $ per MWh

Sliding Scale (Block Rates)
Price either increases or decreases as consumption increases, depending on the intended price signal.
- For volume pricing, as in for large industrial use, the price would decrease as usage increases. This is the usual case.
- For conservation, the price would increase as usage increases.

Market Adjustment
Typical units: $ per kWh
This is used for uncontrolled costs the utility incurs, such as:
- Fuel cost adjustments for generation, and
- Costs for operating peak generating equipment
- Spot purchases due to peaks or unscheduled outings.
- Other unplanned expenses passed on

Time of Use
Pricing signal adjustment for leveling load for the utility.
- Applies to kW and kWh fees.
- On-peak pricing is applied during high demand periods.
- Off-peak pricing is applied in other times.

Demand Charges
Typical Units: $/kW per month or $/kW per day
- Measurement of power draw, not energy use.
- This charge reflects the infrastructure burden on the utility since the instantaneous demand must be met with appropriate generation, transmission, and distribution equipment.

Load Factor
Defined as "average/maximum electric demand."
- Usually not billed directly, but customers with low load factors will end up getting high demand charges.
- Maximum demand is metered and appears on the utility bill. Average demand is calculated:

Avg. Demand = Total kWh/total hours in the period

For example, if a billing period is 30 days long and records 100,000 kWh and 350 kW, find load factor:
Max demand = 350
Avg demand = 100,000/30/24 = 139 kW
Load factor = 139/350 = 40%

Business Type	Average Load Factor
Grocery	75-80%
Health Care	55-65%
Multi-family	50-65%
Retail	50-60%
Lodging	45-60%
Education	45-55%
Restaurant	50-55%
Office	45-55%
Manufacturing—general	15-40%
Manufacturing—semi conductor	85-95%

Figure 6-1. Load Factors by Business Sector

Utility Rate Components

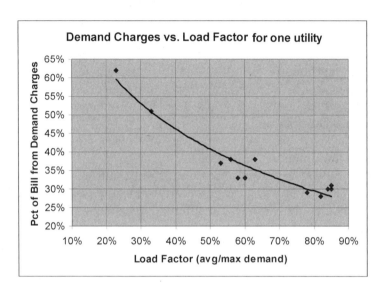

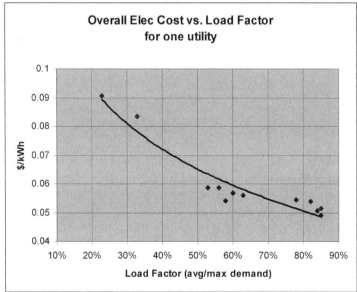

Figure 6-2. Load Factor Relationship to Electric Cost for One Utility

- When a customer uses a high rate of power for a short time, but much less power the rest of the time, the installed utility equipment is underutilized much of the time and represents a stranded cost without sufficient revenue to pay for it. The rate design recovers costs for low load factors to counter the underutilization effect on their revenue stream.
- Load factor is not billed for directly. Rates are designed to increase demand charges for low load factors to recover costs for underutilized equipment.

Power Factor

The difference between "real" and "apparent" power, this factor addresses the magnetizing currents required by electromagnetic devices such as motors, transformers, inverters, and welders, especially when oversized or idling.
- Utility meters record watts which is "real power" and the additional current from power factor less than 1.0 are not recorded on watt meters. Since the added currents are felt by the generators, transmission, and distribution equipment they represent real costs.
- The power factor fee is used to recover these costs. The power factor measurement is usually determined by a gross measurement of kW vs. kVA.

Ratchet Clause
(Minimum Demand)

Typical units are 75% of the highest demand during the prior 12 months.
- Utility rates are set based on assumed utilization of installed equipment. If the customers demand for power decreases suddenly, such as when a large tenant moves out, the installed power supply equipment becomes a stranded investment with insufficient revenue to pay for it.
- The ratchet clause is a common feature in utility rates to recover these costs. The ratchet serves to bill at a minimum demand even when the actual demand is lower.

Interruptible

Utilities that are resource-limited may utilize the interruptible rate as part of their resource supply mix. By turning off a customer's power,

the system capacity is available for other customers without building new generator or distribution capacity.
- In exchange for the ability to interrupt at will, the customer receives lower rates.
- The utility will "call" an interruptible event, at which time the customer agrees to turn off load.
- If the customer fails to turn off load, significant penalties are applied.

GAS

Common utility rate types are summarized here. Infrastructure fees, administrative fees, special customer rates, extremely large customer rates, seasonal, transmission, and de-regulated direct purchase rates are not included.

Energy
Units: therms or dekatherms.
Therm = 10^5 Btu
Dekatherm = 10^6 Btu

In some areas, gas is billed in cubic feet, usually hundreds (ccf) or thousands (mcf). When billed on volume, pressure and temperature corrections apply.

Sliding Scale (Block Rates)
Price either increases or decreases as consumption increases, depending on the intended price signal.
- For volume pricing, as in for large industrial use, the price would decrease as usage increases. This is the usual case.
- For conservation, the price would increase as usage increases.

Market Adjustment
Typical Units: $ per therm or $ per ccf.
- This is used for uncontrolled costs the utility incurs, such as changes in purchase price for gas.

Interruptible

Utilities that are resource-limited may utilize the interruptible rate as part of their resource supply mix. By turning off a customer's gas supply, the system capacity is available for other customers without building new distribution capacity.

- In exchange for the ability to interrupt at will, the customer receives lower rates.
- The utility will "call" an interruptible event, at which time the customer agrees to turn off load.
- If the customer fails to turn off load, significant penalties are applied.

Chapter 7

Automatic Control Strategies

GENERAL

Implementing, expanding, or modernizing an automatic control system, especially when migrating from pneumatic controls, provides a wide array of opportunities for optimization and energy savings.

The cost of a new control system is high. Cost justification for these are usually based on avoided cost of the existing controls if in poor repair, or energy savings opportunities, although often the need for control *at all* is not factored in. This is like trying to find economic justification for having a steering wheel in a car—obviously you need one anyway. The better approach to economic justification is to evaluate the *upgrade* incremental cost between a basic "temperature control only" system and the more high powered optional system or features.

Additional benefits can be realized from improved indoor comfort and reduced response time for service calls; however these are often hard to quantify. The greatest savings from optimization will occur when the customer is willing to embrace new and complex control strategies, however this must be tempered with skill level and training level of operational staff.

The cost of installing a modern control system is considerable, but improvements made with modifications and adjustments to a control system in place can offer among the best returns on investment. It is safe to say that once in place, a modern digital control system can do a lot for energy savings.

COST/BENEFIT RATIO FOR CONTROL SYSTEM ECMS

See **Figure 7-1, Automatic Control System Cost/Benefit Ratios.**

Commercial Energy Auditing Reference Handbook

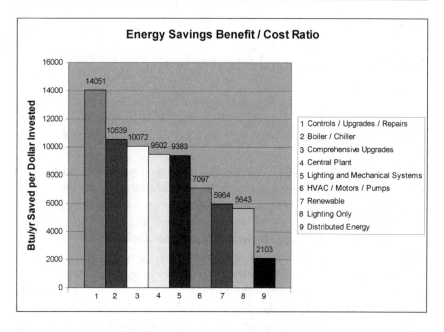

Graph of Linear Data Values
Note: Standard Deviation Ranges are not shown. This is for general comparison only.
Source: "Analysis of Energy Savings in the Federal Sector through Utilities Service Programs," *Strategic Planning for Energy and the Environment*, Vol. 25, No. 4, 2006.

ECM	Btu/yr per dollar	Std. dev
Controls / Upgrades / Repairs	14051	761
Boiler / Chiller	10539	1091
Comprehensive Upgrades	10072	2787
Central Plant	9502	1403
Lighting and Mechanical Systems	9383	511
HVAC / Motors / Pumps	7097	948
Renewable	5964	1241
Lighting Only	5643	270
Distributed Energy	2103	766

Figure 7-1. Automatic Control System Cost/Benefit Ratios

CONTROL SYSTEM APPLICATION NOTES

- For sustained savings, all analog input and output sensors, transducers, and actuators for valves and dampers (travel adjustment and tight close-off) should be re-calibrated on a 2-year cycle, and control set points and occupancy schedules should be reviewed on a 2-year cycle as well.
- A caution in the application of any control system is to assure that the systems under control are "controllable" and that controls are not used in an attempt to solve mechanical design, installation or misapplication issues; always fix those issues concurrently with the new control system.

LIGHTING CONTROL STRATEGIES—BASIC

Programmed Start-Stop
Basis of Savings: Reduced run time.
- Timed coincident with occupant schedules.
- Two features are important for acceptance of these systems:
  — A warning that the lights are about to go out
  — Local overrides that allow people to keep the lights on

Photo Cell Control of Outdoor Lights
Basis of Savings: Reduced run time.
- Outdoor lights off during day-lit hours
- Time switch to shut off some site lighting after dark, but security lighting stays on.

Occupancy Sensor Lighting Control
Basis of Savings: Reduced run time.
- Use for individual offices, meeting rooms, classrooms, multi-purpose rooms, warehouses, etc., that are only occasionally occupied. For warehouses, it is good practice to leave a few lights on continuously for safety and security. Not practical for restrooms, since they can leave people in the dark, unless at least one light is left on.

Timed Tenant Override
Basis of Savings: Reduced run time.
• When after-hours tenants invoke the override, limit the amount of time this will remain active before the system reverts to unoccupied mode again (e.g. 2 hours), and what actually gets overridden.

HVAC CONTROL STRATEGIES—BASIC

Programmed Start-Stop
Basis of Savings: Reduced run time.
• Any equipment left to run continuously should be controlled for automatic start-stop based on occupancy schedules. This can be energy management system (EMS) control or a simple time switch. Outside air dampers are usually closed and exhaust fans off during unoccupied times.
• Settings should closely match actual use. Occupancy settings for worst case may be easiest to implement, but represent lost savings opportunities. For example, if a classroom is normally vacated at 5pm but occasionally occupied until 10pm, setting the occupied controls to always leave it in "occupied" mode until 10pm is wasteful.

Standardize Indoor Comfort Settings
Basis of Savings:
1. Reduced envelope losses by reducing the temperature difference between indoors and outdoors.
2. Standardized temperatures prevent simultaneous heating and cooling from adjacent zones operating at different temperatures.

See **Figure 7-2, Energy-saving HVAC Control Settings.**

Restrict Tenant Adjustment Limits
Basis of Savings:
• Prevent simultaneous heating and cooling from adjacent zones operating at different temperatures.
• Where "zone of greatest demand" optimization routines are used, imposing this limit prevents a rogue zone setting from driving the routine.

Automatic Control Strategies

- Remote lock-out of tenant adjustment for space temperature, or limiting the adjustment to +/- 2 degrees F at most.

Setting	Description	Basis of Savings
75 degF	Space Cooling – Occupied	Compared to lower temperatures: Reduced temperature difference (dT) between inside and outside, lowering envelope transmission. Reduced cooling system lift by raising the evaporator temperature. 1-1.5% per degree efficiency gain in the compressor.
78 degF	Space Cooling – Occupied (demand limiting)	Additional benefit for cooling system during peak demand periods. This may be undesirable if it is outside the comfort envelope.
85 degF	Space Cooling – Unoccupied	Reduced dT between inside and outside, lowering envelope transmission during unoccupied times.
70 degF	Space Heating – Occupied	Compared to higher temperatures: Reduced temperature difference (dT) between inside and outside, lowering envelope transmission.
60 degF	Space Heating – Unoccupied	Reduced dT between inside and outside, lowering envelope transmission during unoccupied times.
5 degF	Minimum dead band between heat and cool	Reduces likeliness of overlapping heating and cooling. Adjacent heating and cooling operations sequenced on a common temperature are very likely to overlap.
1.0 in. w.c.	Duct Static Pressure	Compared to higher pressures: Reduces fan energy. A 10% reduction in duct static reduces fan power by 10% for VAV systems and 15% for constant volume systems. See Chapter 5 - ECM Descriptions **"Lower VAV Duct Static Pressure Control Setting."**
45 degF	Chilled Water Temperature – Dehumidifying	Compared to lower temperatures: Reduced cooling system lift by raising the evaporator temperature. 1-1.5% per degree efficiency gain in the compressor. Note that increasing chilled water temperature above this point will seriously reduce dehumidification potential of air coils, so this should be used with caution in areas where humidity removal is important.
50 degF	Chilled Water Temperature – Sensible Cooling	Compared to lower temperatures: Reduced cooling system lift by raising the evaporator temperature. 1-1.5% per degree efficiency gain in the compressor.

Figure 7-2. Energy-Saving HVAC Control Settings

Continued

Setting	Description	Basis of Savings
(note 2)	Condenser Water Temperature	Compared to higher temperatures: Reduced cooling system lift by lowering the condenser temperature. 1-1.5% per degree efficiency gain in the compressor. Note that there is a balance between compressor savings and added cooling tower horsepower added, as well as the sizing of the cooling tower in "approach" capability. See Chapter 7 - Automatic Control Strategies, HVAC Controls Strategies (advanced) **"Optimum Condenser Water Set Point"**
(note 3)	VAV box cooling minimum air flow	Compared to higher settings: Reduced heating and cooling overlap in areas where demand for air flow is less than allowed by the minimum stop. These areas require supplemental heating or result in over-cooling when internal loads do not match the minimum air flows. This is the "reheat penalty" built into a standard VAV design. Note that the ventilation air in a standard VAV design is mixed in with the supply air, which is the reason for the minimum setting. Reducing this should be done with caution to assure proper ventilation is provided for occupants.
(note 4)	VAV box heating minimum air flow	Compared to higher settings: Reduced reheat penalty. During heating mode, the supply air must be heated to room temperature before any heat added results in heating of the room. The amount of reheat required (the VAV system reheat penalty) depends on the temperature of the supply air and the amount of it. Keeping the heating air flow equal to minimum cooling air flow keeps the reheat penalty to a minimum.

NOTES
1. Cooling comfort settings will vary by climate zone, especially affected by ambient moisture levels. Drier climates will generally allow higher summer temperatures. Equally important to the settings are the separating "deadband" between heating and cooling. This provides a good measure of prevention against simultaneous heating and cooling.
2. Cooling tower performance is almost entirely driven by wet bulb temperature. Since "design" wet bulb temperatures vary by locale, the available condenser water temperature varies as well. A good rule of thumb is wet bulb +7 deg F as the condenser water supply temperature.
3. As low as possible while still meeting general ventilation requirements. This is a complicated topic to fully optimize. The lower the minimum setting, the lower the reheat penalty in winter. For good envelope designs, the conventional VAV system is likely to overcool interior spaces unless supply air reset is employed or outside air is delivered separately and tempered as in the dedicated outside air system (DOAS).
4. Equal to, and no higher than, the cooling minimum setting.

Figure 7-2. Energy-Saving HVAC Control Settings (*Continued*)

Automatic Control Strategies 117

Match Equipment Capacity to Changing Loads
Basis of Savings: Provide "just enough" cooling or heating or air flow or water flow.
- Resetting primary cooling temperatures upward reduces power consumption by 1-1.5% per degree for refrigeration equipment.
- Resetting primary heating temperatures downward, if at the boiler, reduces flue temperature and fuel use by 1% per 40 degrees.
- Resetting primary heating temperatures, if only the distribution system from a mixing valve, reduces standby losses from distribution piping by reducing the differential temperature. Whatever the circulation thermal losses are, they will be reduced proportionally.
- Reducing circulating mass flow rates for air and water reduces circulating fan and pump horsepower proportionally as the reduction of flow (half the flow = half the power).

Vary fluid temperatures, pressures, supply air flows, water flows, outdoor air intake rates, based on demand and not worst case. The objective is always to provide enough, but just enough, of the item.

Optimum Start
Basis of Savings: Reduced equipment run time.
- Use to delay HVAC system start-up as long as possible. During the warm-up mode, the building is normally unoccupied so the outside air damper can remain closed and exhaust fans off until occupied.

Optimum Stop
Basis of Savings: Reduced heating and cooling energy.
- Use to turn off primary heating and cooling equipment (Chillers/Boilers) shortly before the end of the occupied period, utilizing the thermal lag from the circulating fluid, air, building and furniture mass, and the fact that most people will not detect (or complain about) a temperature change of less than 2 degrees F. Ideally, the comfort conditions will have slipped just 2 degrees F just as the occupied period ends.

Occupied/Unoccupied Mode (Set Up/Set Back)
Basis of Savings: Reduced envelope losses by reducing the temperature difference between indoors and outdoors.
- Use to set up and set back space temperatures, and to reduce or stop

outside air intake during unoccupied times. Typical unoccupied temperatures are 55 deg F heating and 85 deg F cooling.
- The need for rapid return to temperature (hotel) or sensitive equipment (computers) may temper the suggested settings.

This model identifies the sequential operational modes. Heating is used for the example; cooling is similar. **See Figure 7-3.**

A. Unoccupied mode begins: unit off, temperature of indoors drifts downward as heat is lost, but no heat is added until the building temperature drops sufficiently to reach the "unoccupied" lowest allowed setting.
B. At unoccupied temperature: At this low temperature, the heating will be allowed to cycle off and on. The hours spent at this low temperature represent the savings, since the reduced inside-to-outside differential temperature (dT) reduces heat transfer and load.
C. Nearing the new occupied period: The heating system must be started in advance of the onset of the new occupied period to allow time for the building to warm up. The heating unit will be on full capacity at this time and may do so for several hours.

Energy use of the ECM is found by piecing together the sequences A,B,C. Variables that may be hard to estimate are:

A. The time, upon entering unoccupied mode, the unit will be completely off.
B. The duration of on and off cycles during the unoccupied period.
C. The time, in advance of the new occupied period, the unit will need to start again—to reach the "occupied" temperature.

Each of these are related and depend on the outdoor temperature: the colder outdoors the quicker the building will drop in temperature to the setback temperature and begin cycling and the sooner it will have to re-start to warm up the building. So the benefit of this measure is dynamic and will vary by building envelope type and climate zone.

Anecdotal estimates of savings vary and actual test data could not be found. A common rule of thumb for set back heating thermostats is:

One (1) percent for each degree of set back that is kept there for at least 8 hours.

Automatic Control Strategies 119

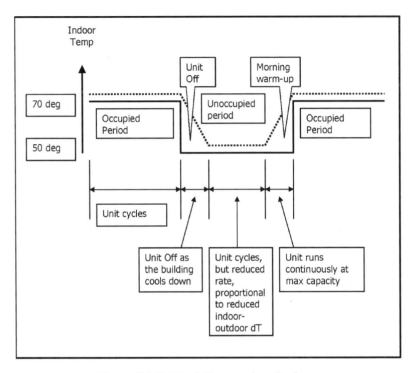

Figure 7-3. Set Back Temperature Savings

Two (2) percent for each degree of permanent change.

A spreadsheet with bin weather for different cities and some assumptions was used to validate the rules of thumb. See **Figure 7-4**.
- Full-off time (A) = 2-4 hours
- Max warm-up time (C) = 1-2 hours
- 10 hour occupied period, 14 hour unoccupied period.
- Bin weather evaluated at 50,40,30,20,10 and 0 degrees OA temp

The analysis verified set back of 20 degrees achieved 33% or better reduction in annual heating energy. Thus, the rule of thumb of 1 percent of envelope heating loss per degree of setback is reasonable and conservative.

SET BACK SAVINGS IN VARIOUS AREAS								
Bin Weather Data, M-F, 6am-6pm								
Saturday 10a-4p								
Sunday off all day							Savings are percent of envelope heating loss	
Point Number	0	1	2	3	4	5	Envelope Savings	
Nominal Value OA	50	40	30	20	10	0	From Setback	
Range	44-54	34-44	24-34	14-24	4-14	[-6]-4	overall	unocc'd period
Colorado Springs, CO	874	962	802	439	111	77	34%	59%
Chicago, IL	768	871	748	411	105	82	34%	58%
Oklahoma City, OK	783	747	402	105	66	10	37%	63%
Tampa, FL	403	130	10	0	0	0	42%	71%
Portland, OR	1633	1115	164	0	0	0	40%	68%
Phoenix, AZ	749	232	13	0	0	0	42%	71%
Anchorage AK	992	923	923	699	342	170	33%	56%

Figure 7-4. Set Back Thermostat Savings
Savings are approximated for a 20 deg F setback.

HVAC Occupancy Sensor Control

Basis of Savings: Reduced equipment run time, reduced air flow.

- The same concept of lighting control via occupancy sensors can be used for HVAC.
- Intermittently occupied rooms or areas, such as hotel rooms, meeting rooms, or classrooms, are candidates for this. When it is determined that the room is unoccupied, the occupancy sensor input to the room controller resets the temperature to unoccupied levels.
- Additionally, for VAV systems, the unoccupied mode sensor resets the VAV box minimum setting to zero or a modified reduced minimum if near a perimeter wall or glass. When the room temperature goes outside the unoccupied zone, the minimum stop of the VAV box is set back up to the original minimum.

Timed Tenant Override

Basis of Savings: Reduced run time.

- When after-hours tenants invoke the override for comfort, limit the amount of time this will remain active before the system reverts to unoccupied mode again (e.g. 2 hours), and what equipment needs to operate.

Deadband Thermostats

Basis of Savings: Prevent simultaneous heating and cooling.

- Use or adjust thermostats to allow for a zero-energy deadband, a range where no action is taken by the HVAC system. A simple, but

Automatic Control Strategies

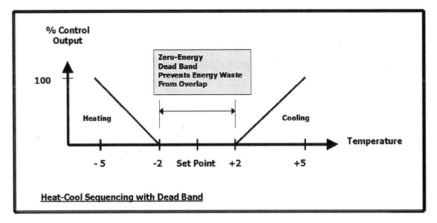

Figure 7-5. Diagram of Deadband Control
Source: Energy Management Handbook, 6th Ed, Chap 22, Turner/Doty.

effective routine that should be part of every non-critical comfort control application. The deadband separation of heating and cooling controlled devices provides a positive separation to prevent the energy waste associated with heat/cool overlap.
- A minimum of four (4) deg F deadband for space temperatures is suggested.
- A minimum of five (5) deg F deadband for sequential air handler heating/cooling equipment is suggested.

VAV Box Control Deadband
Basis of Savings: Prevent simultaneous heating and cooling.
- Independent VAV heating and cooling set points with a minimum 5 degree zero-energy deadband in between, in lieu of a single setting with a proportional band. This will increase the zero-energy deadband and reduce simultaneous heating and cooling.

NOTE: For normal VAV systems where the ventilation air comes mixed with the supply air, there is a minimum stop setting, for minimum ventilation, that creates some heating-cooling overlap. The deadband application for VAV boxes minimizes, but does not eliminate this overlap.

Air Handler Control Deadband
Basis of Savings: Prevent simultaneous heating and cooling.
- Independent preheat, mixed air, and cooling coil control set points, with a minimum five (5) degree zero-energy deadband in between,

in lieu of a single setting with a proportional band. This will reduce simultaneous heating and cooling.

Shed Non-essential Equipment at Peak Times
Basis of Savings: Savings are from reduction in utility demand charges as well as energy savings.
- Schedule things like fire pump testing and battery charging to off peak times.
- Any load shifting to off peak serves this measure.
- Demand charges are usually determined from the greatest demand recorded any time in the month, so this must be done faithfully to achieve savings, i.e. through automation.

Load Limiting
Basis of Savings: Savings are from reduction in utility demand charges as well as energy savings.
- Prior to beginning of and through the duration of identified peak times.
- Reducing demand for primary electrical HVAC equipment can save money on demand costs. This has application for chillers, large variable speed fans, and large variable speed pumps. For example, by load limiting centrifugal equipment like chillers, pumps, and fans to 90%, this keeps 90% of design capacity available but will reduce electric demand by 20-25%. Since 30-40% of the electric costs for large facilities are often from demand, this will create savings.
- Demand charges are usually determined from the greatest demand recorded any time in the month, so this must be done faithfully to achieve savings, i.e. through automation.

Outside Air for Morning Building Cool-down
Basis of Savings: Reduced heating and cooling energy related to tempering outside air.
- Use outside air for pre-occupancy cool-down cycle in the summer, in conjunction with scheduled off time and/or optimal start routines. This routine takes advantage of cool nights and flushes the warm air out of the building that has accumulated and avoids or reduces demand on the primary cooling equipment.

NOTE: Normally limited to dry climates. In humid climates or whenever the outside air humidity is higher than building humidity, this has the potential to add load if subsequent de-humidification (and energy expenditure) occurs as a result.

Extended Full Economizer Range of Operation
Basis of Savings: Reduced run hours of cooling compressor equipment.
- Adjust settings for economizer operation based on outdoor dew point levels to keep cooling equipment off longer during mild and dry weather days.
- If outside air dew point levels are below 47 degrees and outside air dry bulb temperature is below 65 degrees, it should be possible in most cases to circulate the available air instead of 55 degree air and maintain comfort in many buildings. Doing this will delay the use of mechanical cooling and create savings.
- Much of the savings in mechanical cooling energy is taken by increased air flow rates in VAV systems, however there is still a net gain. Rule of thumb is that half of the savings from keeping the compressors off is spent in added fan energy.
- Some form of humidity watch-dog control is suggested except in very dry climates.

Extended Partial Economizer Operation
Basis of Savings: Reduced mechanical cooling energy use in periods where full economizer is not practical but outside air in combination with mechanical cooling is more economical than plain mechanical cooling.
- Extend air-side economizer operation as far as practical, with the use of enthalpy comparison sensors, in lieu of the standard practice of limiting the economizer operation to 55 degrees F or less. With this method, whenever the outside air has less energy in it than the return air, cooling costs will be reduced by using outside air instead of return air. On days when the outside air is significantly drier than inside air, the free cooling range can be extended and outside air temperatures above return air temperatures can be utilized. When the outside air is humid, 55 degrees F is a smart cutoff point.

NOTE: Use this method with caution. Enthalpy sensors are difficult to calibrate and so accuracy of the readings and resulting control decisions have a measure of uncertainty with them. Actual results may vary from predicted results depending on accuracy and drift. Comfort complaints will be the feedback to operations if they drift in one direction, but lost savings will be the result of drifting in the other direction, with no feedback.

Boiler Lockout from Outside Air Temperature
Basis of Savings: Reduced run time of the boiler.
- Generally, prohibit boiler operation above 60 degrees F. Suggested control would be to cut-in at 55 and cut-out at 60 degrees F outside air. This is open loop control, but saves energy compared to "start based on demand" routines, since all it takes is one strong demand point to run the large machine for extended periods, or to start it unnecessarily in hot weather.

Chiller Lockout from Outside Air Temperature
Basis of Savings: Reduced run time of the chiller.
- Generally, prohibit chiller operation below 50 degrees F, and coordinate with air and water economizer settings. Suggested control would be to 'cut-in' at 55 and 'cut-out' at 50 degrees F outside air. This is open loop control, but saves energy compared to "start based on demand" routines, since all it takes is one strong demand point to run the large machine for extended periods, or to start it unnecessarily in cold weather.

Chilled Water Reset—Constant Flow Pumping
Basis of Savings: Improved refrigeration cycle.
- Approximately 1-1.5% power reduction for each degree raised.
- Reset is from return temperature. As cooling load is reduced, return water temperature drops and reflects the reduced load.
- Chilled water supply temperature is raised according to a reset schedule. Example reset schedule:

CHW Return	CHW Supply
55 degF	45 degF
50 degF	48 degF

- Limits must be placed on the reset schedule. Chilled water temperatures above 50 degrees will do very little comfort cooling and can significantly reduce dehumidification action in cooling coils.

Chilled Water Reset—Variable Flow Pumping
Basis of Savings: Improved refrigeration cycle.
- Approximately 1-1.5% power reduction for each degree raised.

- Raising chilled water temperature increases flow rate requirements per ton and increases pump energy, eroding savings; usually eroding half to 2/3 of the savings. Still a net gain.
- Reset cannot be from return temperature since return is relatively constant with this system.
- Cooling load reduction is either measured directly (tons) or implied by outside air temperature.
- Chilled water supply temperature is raised according to a reset schedule. Example reset schedules:

Cooling Load	CHW Supply
100%	45 degF
50%	48 degF

Outside Air	CHW Supply
80 degF	45 degF
60 degF	48 degF

- Limits must be placed on the reset schedule. Chilled water temperatures above 50 degrees will do very little comfort cooling and can significantly reduce dehumidification action in cooling coils.

Outside Air Reset of Hot Water Converters
Basis of Savings: Reduced standby losses from heating fluid distribution piping.
- Compared to a constant temperature setting, this will reduce standby losses. In the cooling season, these heat losses are also unnecessary cooling loads.

Mid-Range Vestibule Temperature
Basis of Savings: Reduces temperature difference at the envelope opening means less loss.
- For vestibule spaces, temper the space to a mid-range temperature, half way between indoor and outdoor temperatures.

Operable Window Interlock
Basis of Savings: Reduce loss of conditioned air whenever windows are opened. Lost air results in more make-up air that must be tempered.

- Where operable windows are used, provide automatic control interlock to disable the HVAC serving that room, to avoid heating and cooling the outdoors.

Roll-Up Door Interlock
Basis of Savings: Reduce loss of conditioned air whenever windows are opened. Lost air results in more make-up air that must be tempered.
- Provide automatic control interlock to disable the HVAC serving interior area, to avoid heating and cooling losses when the door is up. This will encourage people to keep the door closed.

Staggered Heating and Cooling System Start-Up
(Buildings with demand charges)
Basis of Savings: Reduced demand charges by avoiding setting a peak event.
- After extended 'off' periods, such as night set back, normal automatic control response will be to drive the heating and cooling equipment in order to reach the occupied set point. To avoid setting the utility maximum electrical demand value and subsequent higher demand charges coming out of unoccupied periods, bring large electrically driven heating and cooling equipment loads on in segments for 30 minutes or until the process has "caught up" and is no longer at full load.

NOTE: This is not in reference to inrush currents for starting motors.

For example:
Consider a building heated by electric resistance heat. By default, coming out of unoccupied periods during winter will result in 100% of the heat being energized simultaneously. Controlling in 2 or more zones and allowing ample pull-up time for zone 1 to get to temperature and begin cycling normally before starting zone 2 will reduce the maximum demand for that day and, if done diligently through automation, will reduce the overall seasonal demand and demand charges.

- Electric resistance heat (winter)
- Packaged HVAC cooling equipment (summer)

LIGHTING CONTROL STRATEGIES—ADVANCED

Daylight Harvesting
Basis of Savings: Reduced run time of electric lighting.
- Control perimeter lighting on/off or modulate, in response to daytime sunlight entering the building.
- In-board/out-board switching for interior lighting.

Programmable Lighting Ballast
Basis of Savings: Reduced run time with additional flexibility in scheduled operation.
- Special "addressable" lighting ballast are available that can be controlled individually, without conventional relays. Remote control, custom scheduling by room, by area, and by light fixture are possible with this system to minimize lighting use.

HVAC CONTROL STRATEGIES—ADVANCED

NOTE: Many optimization routines rely on end-use polling of demand or valve/damper positions such as "most open valve" routines. This can be done cost effectively without actual measurement of position—by polling the individual "percent commanded output." This is referred to as "implied position" and is acceptable in most cases in lieu of actual position.

Note also that routines that use polling have the potential to be inefficiently operated if one errant measurement exists. For polling space temperatures, for example, limiting the user adjustment is strongly recommended in conjunction with demand polling of space controls. Additionally, it may make sense to 'discard' the high and low values from such polling to prevent errant operation. Some polling techniques wait to react until several "calls" exist; this reduces the chance of an errant signal driving the entire heat/cool plant, but also introduces dissatisfaction if a single and legitimate call exists, since the control system would ignore it.

Occupancy Sensor Control of HVAC
(where a dedicated VAV box or terminal unit exists)
Basis of Savings: Reduced heating and cooling load. Reduced VAV system reheat penalty.
- When the room is sensed as unoccupied, the space temperatures revert to unoccupied values. For VAV boxes, minimum flow settings are adjusted to zero.

NOTE: Perimeter areas strongly influenced by envelope loads may need specially defined "day time unoccupied" settings to avoid comfort complaints and to allow quicker return toe occupied temperatures when the room is re-occupied.

Optimized VAV Box Minimum Setting Reset
Basis of Savings: Reduced VAV system reheat penalty.
- The usual method of calculating VAV box minimums is to assume full load conditions and the outside air percent of the total supply air at those conditions. With constant outside air control, the mixture is richer in outside air at supply air flows that are below maximum. Therefore, the VAV minimum stops can be reset automatically based on the actual OA mix as loads change.

Example:
Assume a 20% VAV minimum stop is required based on 25% OA at design conditions. This same VAV box could have a minimum position of 10% when the supply fan is at 50%, since there is now twice as much OA in that air stream. The effect of this is to reduce over-cooling and reheating when the VAV box is at minimum while the zone thermostat continues to call for less air.

Optimized Supply Air Static Pressure Reset—VAV Systems
Basis of Savings: Reduced fan horsepower via affinity laws.
- This requires polling of individual VAV boxes for air valve position and reduces system pressure until at least one box is 90% open, thereby providing the optimal system duct pressure (just enough pressure). This reduces fan horsepower.

Optimized Supply Water Pressure Reset—Variable Pumping Systems
Basis of Savings: Reduced pump horsepower via affinity laws.

Automatic Control Strategies 129

- This requires polling individual air handlers for control valve position and reduces system pressure until at least one control valve is 90% open, thereby providing the optimal system water pressure (just enough pressure). This reduces pump horsepower.

Condenser Water Reset and Optimized Evaporative Cooling Setting
Basis of Savings:
1. *Cooling towers: refrigeration compressor power savings of 1-1.5% per degree lowered. Optimizing allows this to occur without excessive tower fan energy penalty.*
2. *Evaporative cooling sequenced with mechanical cooling: reduced run hours of mechanical cooling equipment, while avoiding moisture issues from excessive indoor humidity.*

- This uses outdoor air wet bulb temperature, which can be measured directly, but is usually calculated from temperature and humidity. The evaporative process can get close to, but never reach or exceed, the wet bulb temperature. By knowing the wet bulb temperature, the control system will know its boundaries and won't try to achieve something it cannot. For water-cooled refrigeration equipment, the low limit for the reset is normally around 55 or 65 degrees F and the chiller manufacturer needs to be consulted to confirm. Resetting the cooling water temperature down in this way, in lieu of a constant temperature setting, will reduce kW/ton energy use and demand. NOTE: achieving colder condenser water from a cooling tower is a trade off between improved chiller kW/ton and increased cooling tower fan kW/ton, so evaluation for diminishing returns is required.

- See **Chapter 9—Quantifying Savings for expanded example of this ECM**

**Optimized Supply Air Temperature Reset
for VAV Air Systems from Zone Demand**
Basis of Savings: Reduced VAV reheat penalty in winter without losing cooling system benefit of constant cold air temperature.
- This requires polling individual VAV box air valve position and reheat valve position. This utilizes a fixed temperature for cooling (e.g. 55 degrees F) with no reset at all until polling of individual boxes in-

dicates that most of the boxes are in heating—only then is the SA temperature allowed to gradually be reset upward to its maximum limit (e.g. 62 degrees F). The reset is accomplished by polling VAV reheat valve positions and the air temperature is gradually reset upwards until at least one VAV box reheat valve is 90% open, thereby providing optimal air temperature (just warm enough). Simultaneously, VAV damper positions are polled to be sure enough cooling is being provided for any zone still calling for cooling, and cooling will usually prevail if the two are at odds.

- A similar, but less accurate, method uses the supply fan VAV output percent command along with outside air temperature to estimate when most of the VAV boxes are at minimum and in the heating mode, signaling the time to begin resetting supply air temperature. An important feature of this reset schedule is that NO reset occurs during summer months, preserving the basis of VAV savings. A simple reset schedule is used for this method as follows:

SA Fan % Capacity	Outside Air Temp	Supply Air Temp Set point
100%	Any	55
40%*	40†	55
20%	10	62

*The percent of the supply fan capacity at which the reset begins should correspond to the aggregate VAV box minimum settings. For example, if the aggregate (weighed average) VAV box setting is 40%, then a supply fan value of 40% indicates that most, but not all, VAV boxes are at minimum.
†The temperature at which the reset begins should correspond to the building thermal break-even point, which is the point below which the 'self heating' nature of the building is overcome by envelope losses and ventilation loads and heating use begin to dominate.

NOTE: this method will require some experimentation to get right. Percent fan capacity (VFD output, inlet vane position, etc.) is close, but not equal, to percent of air flow.

- Another method for supply air reset of VAV systems uses outside air temperature to imply when heating loads will begin to appear. This is easy to implement, but is an open loop control system, i.e. no amount of change in supply temperature can alter outside temperature. A reset schedule for this method would look like the following:

Outside Air Temp	Supply Air Temp Set point
Above 70	55
60 *	60
Below 60 *	60

*The values of the lower reset parameter will vary depending on the break even temperature.

Demand Controlled Ventilation (DCV)/Ventilation Reset from CO_2
Basis of Savings: Less outside air to heat and cool.
- Has application for large open assembly areas, characterized by a single point of CO_2 control in the occupied space of the single zone unit serving that area. Applications include a ballroom, theatre or open plan office area. A separate CO_2 point of control would be required for each dividable meeting area, each assembly area, and each group area, so that areas of great ventilation demand are served appropriately and do not get overlooked by sensor averaging.
- In essence, this control method uses a CO_2 sensor(s) as a people counter, thereby optimizing the use of outside air, and the energy required to condition it.

Optimized Ventilation Reset by People Count
Basis of Savings: Less outside air to heat and cool.
- A combination of number of people and the duration at each location would provide the necessary information to calculate the exact amount of ventilation air needed. This method puts the welfare of the occupant first, but also minimizes the energy use by providing the optimal ventilation quantity (just enough ventilation) at any given time.

NOTE: This routine is conceptual only, but is presented due to the large potential for energy savings. To the author's knowledge, the friendly "people counter" instrument, other than a turnstile, hasn't been developed yet. This may be possible, manually or automatically, based on ticket sales for assembly occupancies.

Optimized Ventilation Effectiveness by Season
Basis of Savings: Less outside air to heat and cool.
- Adjust ventilation rates based on improved "e" (ventilation effectiveness) in summer (value 1.0) compared to winter (value 0.8), to save outside air conditioning energy costs.
- This assumes ceiling diffusers for air distribution, whereby the warm air in winter tends to hug the ceiling and the entrained ventilation air doesn't fully reach the breathing zone—hence the 0.8 factor in heating mode. Assuming the ventilation calculations are penalized for 'worst case' winter conditions, then suitable ventilation would be achieved in cooling mode with 25% less ventilation.

Schedule Ventilation Rates to Correspond with Planned Occupancy Patterns
Basis of Savings: Less outside air to heat and cool.
- Application is different occuupancy for different shifts of workers.

Optimized Sequencing of Multiple Chillers/Boilers
Basis of Savings: Efficiency differential of better choice equipment vs. default choice equipment.
- Application is different occupancy for different shifts of workers.
- Strategically selecting cut-in and cut-out points to keep the primary equipment operating in its most efficient range. Typically, maintaining this equipment between 50-90% load achieves good efficiency, but verifying the actual best-efficiency points or range for each system, from manufacturer's load profile data, can provide additional savings.

Optimized Hot Water Reset from Zone Demand
Basis of Savings: Reduced standby loss of circulating heated fluid.
- This requires polling of individual hot water points of use, for control valve position, such that at least one valve is open 90%. By resetting based on demand, this routine will provide the optimal water temperature (just hot enough).

Multi-zone Hot Deck/Cold Deck Reset from Zone Demand
Basis of Savings: Reduced simultaneous heating and cooling, which is inherent in this HVAC system type.
- This requires polling of individual multi-zone (MZ) mixing damp-

ers to determine the greatest cooling and heating demands, ideally so that at least one zone damper is 90% open to cooling and another zone damper is 90% open to heating. By resetting hot and cold deck from space demand, optimal heating and cooling (just enough of each) will be provided in the hot and cold decks. Since these systems inherently mix cooled and warmed air, this reduces simultaneous heating and cooling.

Dual Duct Terminal Unit: Split Damper and Add Deadband
Basis of Savings: Reduced simultaneous heating and cooling, which is inherent in this HVAC system type.
- Unless constant volume is critical to the air system design, splitting the two dampers for independent control allows optimization. With the dampers split (separate actuators), and ventilation air in one or the other air streams (usually the cold duct), control like a VAV box with a deadband. During a call for cooling, the hot duct damper would remain closed while the cold duct damper modulates to maintain temperature. As cooling demand decreases, the cold duct damper throttles toward closed and finally reaches its minimum position (for ventilation). Here it will float within the deadband. If space temperature falls sufficiently to require heating, the hot duct damper would begin to throttle open with the cold duct damper at minimum. This sequence minimizes heating-cooling overlap and significantly reduces the energy use of the dual duct system that otherwise has simultaneous heating and cooling waste built into it.

District Heating and District Cooling Delta-T Control
Basis of Savings: Reduced pump horsepower from reduced flow that comes from high differential temperatures. Savings via affinity laws.
- By actively controlling differential temperatures through buildings points of use to be as high has practical, flow can be reduced and pumping costs minimized. This can be touchy and, if too aggressive, can lead to comfort issues.
- Many, if not all, of these distribution systems pump more water than they need to. There are a variety of reasons for this, which are beyond the scope of this report, but many of them can be mitigated by imposing a control limit of minimum delta T (differential temperature) across the coil, building, or segment of the distribution system.

This has the effect of requiring the point of use to extract all the available energy out of the circulated fluid before returning it and thus reduce pumping volumes and pumping energy.
- Control is implemented by some manner of throttling device, either a control valve or a pump speed, and supervised by differential temperature measurement and/or flow measurement.
- Applications vary, but the common theme is to wring the heat out of the water and reduce flow when possible, while still maintaining comfort.

OTHER CONTROL SEQUENCES

Water Side Economizer
Basis of Savings: Maximizing the use of the flat plate heat exchanger system.
This is the conventional case of chilled water cooling needs being provided by either the chiller or by the water-side economizer system through evaporative cooling and a heat exchanger or fluid cooler, where chiller and water-side economizer modes do not run concurrently.

*Water-side economizer **cut-in** logic:*
Whenever the outside air wet bulb temperature is more than (10) degrees (adjustable) lower than the chilled water temperature set point, chilled water will be provided by the water-side economizer and the chiller will be off.

*Water-side economizer **cut-out** logic:*
Option 1
Whenever the outside air wet bulb temperature is less than (5) degrees (adjustable) lower than the chilled water temperature set point, chilled water will be provided by the chiller and the water-side economizer will be disabled.

Option 2 (extended operation)
Once engaged, the water-side economizer will remain active until the chilled water temperature has exceeded the set point by two degrees (adjustable), after which time chilled water will be provided by the chiller and the water-side economizer will be disabled.

Automatic Control Strategies

Change-over between water-side economizer and chilled water cooling (water-cooled chillers):

NOTE: This is not an energy feature but is such a common operations complaint that a standard control sequence to mitigate the difficulty and encourage acceptance. The issue is switching back to regular cooling mode, when the condenser water has been used as chilled water and is so cold that the chiller may not start when this cold water is re-applied as condenser water.

During water-side economizer operation, the cooling tower operating set point is lowered to a point sufficiently below chilled water temperature to create the necessary heat exchange to cool the building. Upon reverting to normal chiller operation, the chiller cooling water will therefore be much colder than normal and make starting the chiller difficult. To mitigate this, upon initial start-up while leaving the water-side economizer mode, the water flow through the chiller will be throttled (control valve or pump speed controller) initially to one fourth of its normal flow and then modulated to maintain an average condenser temperature (outlet—inlet) of 70 degrees, until such time as the entering condenser water temperature exceeds 65 degrees, at which time the condenser flow will be adjusted to achieve normal design water flows and no further throttling of condenser water will be performed.

OTHER WAYS TO LEVERAGE DDC CONTROLS

- Utility load tracking and real-time feedback to proposed efficiency changes; early warning of high demand.
- Predictive maintenance for heat exchangers (fouling) from measuring approach temperatures, filter changing (pressure drop), etc.
- Global point sharing that justifies higher quality instrumentation, such as outdoor air temperature, humidity, dew point, wet bulb, etc.

Chapter 8
Building Operations and Maintenance

FACILITY REPAIR COSTS

Setting aside a capital reserve of 2% to 3% of the total cost of the building will usually be adequate to replace major systems (equipment, roof) at the end of their life. Without such planning, many building owners will experience large unplanned expenses around 20-25 years.

MAINTENANCE VALUE

Improved HVAC preventive maintenance resulting directly from technician training has a significant energy savings potential. From a study of nine community colleges in California, savings potential estimates ranged from 6% to 19% of total annual campus energy costs, or $0.09-0.26/SF-yr. [1999 dollars]

Source: "Quantifying The Energy Benefits of HVAC Maintenance Training and Preventive Maintenance," *Energy Engineering*; Vol. 96; Issue 2, 1999.

POOR INDOOR COMFORT AND INDOOR AIR QUALITY COSTS

Comfort Productivity Increase
"Impaired Air and Thermal Quality": 1.5%

Source: NEMI, National Energy Management Institute, Productivity Benefits Due to Improved Indoor Air Quality, August 1995, pp. 4-9.

Indoor Air Quality Productivity Increase
Unhealthy Building: 3.5%
("Conditions similar to an SBS building ...but with a lower percentage of employees affected.")

138   Commercial Energy Auditing Reference Handbook

Source: NEMI, National Energy Management Institute, Productivity Benefits Due to Improved Indoor Air Quality, August 1995, pp 4-9.

Indoor Air Quality: 6% for SBS/BRI
[Sick Building Syndrome/Building-related Illness]

Source: NEMI, National Energy Management Institute, Productivity Benefits Due to Improved Indoor Air Quality, August 1995, pp-4-9.

To identify a dollar value of improvements to comfort or IAQ, use the appropriate percentage productivity increase with the total productivity benchmark value.

PRODUCTIVITY VALUE

$/SF-yr	Building Use
$13	Assembly
$45	Education
$19	Food Service
$110	Healthcare
$19	Lodging
$23	Mercantile and Services
$97	Office

Figure 8-1. Worker productivity costs $/SF [1995 dollars]
Source: Table data derived from gathered data in the following document: NEMI, National Energy Management Institute, Productivity Benefits Due to Improved Indoor Air Quality, August 1995, pp 4-8, 4-11.

For example, if a 1% productivity increase (from improved comfort or IAQ) is expected for a 25,000 SF office building with a total productivity rate of $97 per SF-yr, the savings would be 0.01*$97*25,000 or approximately $24,250 per year.

NOTE: Claims of productivity increases equated to dollars are often not received well since they are difficult to measure. It is suggested that attempts to quantify these benefits only be used in cases of extremely poor comfort

Building Operations and Maintenance

or indoor air quality, when the owner has specifically asked for improvements in these areas. It is also suggested to de-rate the results (e.g. multiply by 0.7 or 0.5) to add further credibility to the claim.

MAINTENANCE ENERGY BENEFITS

Cleaning Evaporator Air Coils
- Up to 20% energy penalty in hp per ton, if badly fouled.
- Retarded heat transfer requires greater differential temperature, and lower suction pressures, with corresponding loss of efficiency and extended run times.

Cleaning Condenser Coils
- Up to 20% energy penalty in hp per ton, if badly fouled.
- Retarded heat transfer requires greater differential temperature, and higher head pressures, with corresponding loss of efficiency and extended run times.

Cleaning Condenser Coils and Evaporator Coils
- Individual efficiency losses of condenser and evaporator are additive if both coils are badly fouled.

	Reduction in Capacity	Pct. Increase in Energy Use, HP Per Ton
Dirty Condenser	8.2	20
Dirty Evaporator	18.9	18
Dirty Condenser and Evaporator	25.4	39

Figure 8-2. Effect of Dirty Coils on Energy Use
Data are for a 15-ton reciprocating compressor. Source: *Handbook of Energy Engineering*, 5th Ed, Thumann/Mehta.

Early Filter Change Out
- The average between initial and final pressure drops equal the amount of average fan resistance and horsepower associated with

filters. Some customers delay filter changes to the maximum differential pressure (dP) allowed by the filter manufacturer. While this reduces filter costs, it increases fan energy cost. Lowering the change-out point lowers the average dP and reduces fan energy.

NOTE: Results are predictable for VAV systems, where air flow requirements are determined from thermal demand. For constant volume systems, the effect is negligible unless the fans are re-balanced because greater cfm will result from reduced friction.

Cleaning Chiller Condenser Tubes
- Typical savings are 1-1.5% hp per ton improvement for each degree the refrigerant condensing temperature is lowered.
- Retarded heat transfer requires greater differential temperature, and higher head pressures, with corresponding loss of efficiency.

Cleaning Chiller Evaporator Tubes
- Typical savings are 1-1.5% hp per ton improvement for each degree the refrigerant suction temperature is raised.
- Retarded heat transfer requires greater differential temperature, and lower suction pressures, with corresponding loss of efficiency.
- Most chilled water systems are "sealed" and fouling is minimal. This measure applies mostly to un-sealed systems or possibly to a sealed system with chronic leaks and long term use of make-up water.

Cleaning Chiller Condenser Tubes AND Evaporator Tubes
- Individual efficiency losses of evaporator and condenser tubes are additive if both tubes are badly fouled.

Cleaning Boiler Fire Tubes
- Savings is a function of soot thickness.
- Retarded heat transfer requires greater differential temperature, and higher combustion temperature, resulting in higher stack temperatures, with corresponding loss of efficiency.

Cleaning Boiler Water Tubes
- Savings is a function of mineral build-up thickness and type of mineral.
- Retarded heat transfer requires greater differential temperature, and higher combustion temperature, resulting in higher stack temperatures, with corresponding loss of efficiency.

Losses Due to Soot Build-up in a Boiler

Soot Layer on Heating Surfaces, inches	Increase in Fuel Consumption (%)
1/32	2.5
1/16	4.4
1/8	8.5

Losses Due to Water-side Fouling in a Boiler
(% Increase in Fuel Consumption

Scale Thickness, inches	Scale Type		
	"Normal"	High Iron	Iron Plus Silica
1/64	1.0	1.6	3.5
1/32	2.0	3.1	7.0
3/64	3.0	4.7	—
1/16	3.9	6.2	—

Note: "Normal" scale is usually encountered in low-pressure applications. The high iron and iron plus silica scale composition results from high-pressure service conditions.
*Extracted from National Institute of Standards and Technology, Handbook 115, Supplement 1.
Source: "Clean Boiler Water-side Heat Transfer Surfaces," Industrial Technologies Program (ITP) Steam Tip Sheet #7, December 1999, US DOE Office of Energy Efficiency and Renewable Energy.

Figure 8-3. Boiler Fouling Losses
Source: American Boiler Manufacturers Association.

Cleaning Boiler Fire Tubes AND Water Tubes

- Individual efficiency losses of boiler tubes and fire tubes are additive if both tubes are badly fouled.
- Monitor heat exchange equipment for "approach temperatures."
- By knowing the baseline approach performance, the actual measured approach will infer heat exchange fouling and provide the prompt for predictive maintenance. As the approach temperatures creep up, this usually indicates fouling.

- The very best indication of "clean, new condition" approach is from start-up testing data, or immediately following a thorough cleaning event. As a general starting point, the following are typical values of approach temperatures.
- NOTE: In some cases, reduced air or water flow through the heat exchanger can cause readings that look like fouling but are not.
- NOTE: For equipment with variable loads, the approach temperature will vary somewhat with load. Knowing baseline approach values for 100%, 75%, and 50% load will allow good predictions even at part load.

The rule of thumb used is a 1% penalty in efficiency for each 40 degree F rise in stack temperature above design from fouling. The higher the excess air, the higher the factor since the excess air is cooling the flue gas, so a 40-degree rise requires more heat loss.

Efficiency Gains from Cleaning Fouled Combustion Heat Exchangers

Typical industry rule of thumb for this is 1 degree per 40-degree reduction in stack temperature. However, this value will be higher for burners with high excess air.

Excess Air	Pct improved per 40 deg F reduction in stack temperature
10%	1.2%
20%	1.3%
30%	1.4%
40%	1.5%
50%	1.6%
60%	1.8%
70%	1.9%
80%	2.0%
90%	2.1%
100%	2.2%

Figure 8-4. Efficiency Gains from Cleaning Fouled Combustion Heat Exchangers

Heat Exchange Arrangement	Between Where and Where	Typical Value	Diagram
Counter-Flow Heat Exchanger, General Case, Water or Air	Heating: Hot side in minus cold side out Cooling: Hot side out minus cold side in	---	Heating: A Cooling: B
Parallel Flow Heat Exchanger, General Case, Water or Air	Heating: Hot side out minus cold side out Cooling: Hot side out minus cold side out	---	Heating: C Cooling: D
Water-Cooled Condenser (shell and tube)	Saturated condensing temp minus leaving condenser water temp. This can be approximated by the liquid temperature	0.5-5	E
Water-Chiller Evaporator (shell and tube)	Leaving chilled water temp minus saturated evaporator (suction) temp	0.5-5	F
Air-Cooled Condenser	Saturated condensing temp minus entering ambient air temp. This can be approximated by the liquid temperature	25-40	G
DX Cooling Coil	Leaving air temp minus saturated evaporator (suction) temp	15-30	H
Dry Cooler	Fluid out temp minus entering air temp (ambient)	30	I
Hot Water Boiler	Flue gas out temp minus leaving hot water temp	75-150	J
Fired Water Heater	Flue gas out temp minus leaving hot water temp	20-100	J
Fired Steam Boiler	Flue gas out temp minus saturated steam outlet temp	75-150	K
Fired Air Heating Furnace	Flue gas out temp minus leaving air temp	20-100	L
Steam Heater	Saturated steam temp minus leaving hot water temp	10-30	M
Cooling Tower	Leaving (sump) water temp minus ambient wet bulb temp	7-15	N
Fluid Cooler (coil pack)	Leaving process fluid temp minus sump water temp sprayed onto the coil pack	10-20	P
Chilled Water Coil – Air Cooling (counterflow multi-row coil, coldest air in contact with coldest water)	Leaving air temp. minus chilled water inlet temp	7-10	Q
Hot Water Coil – Air Heating (counterflow multi-row coil, hottest air in contact with hottest water. 50 degF approach is for single row coils)	Hot water supply inlet temp minus leaving air temp	10-50	R
Shell and Tube – Heating, Water-to-Water, Hottest Water in the Shell	Shell water outlet temp minus tube outlet temp	10-20	S
Shell and Tube – Heating, Water-to-Water, Hottest Water in the Tubes	Tube water outlet temp minus shell water outlet temp	10-20	T

Figure 8-5. Heat Exchanger Approach Values

Notes on Heat Exchangers

The measure of heat exchanger performance is "approach" which generally means the temperature difference that can be achieved by the device to heat or cool some gas or fluid with a supply or ambient gas or fluid. The approach is strongly influenced by heat transfer surface area and turbulence at the boundary layers. For the infinitely sized heat exchanger, the leaving fluid can be an exact match for the ambient fluid supplied, but in practical terms it can only *approach* it.

The **shell and tube** heat exchanger in general has a fundamental limitation such that the two leaving fluid temperatures can only approach each other. For most of these types of heat exchangers, the two temperatures can only be economically brought to within 5 or 10 degrees of each other, and 10 degrees F is typical. For special applications of the shell and tube heat exchanger, such as the refrigerant condenser, special baffles can put the liquid refrigerant intimately in contact with the entering fluid for sub-cooling of the liquid after the bulk of the heat transfer has occurred in the main body of the shell.

The **plate-frame** heat exchanger, by its nature, is able to achieve much closer approach temperatures and can actually 'cross' temperatures such that the leaving fluid of one stream approaches the entering (not leaving) temperature of the other stream. This is a special application of the normal concept of "approach" temperatures when discussing heat exchangers, and is also fundamental efficiency advantage that plate frame units have that should be pointed out. Plate frame heat exchangers are also very compact. However there are other considerations of plate frame heat exchangers such as cleaning and first cost. For example, high pressure variations of these units include brazing of the plates to increase pressure ratings (e.g. for refrigeration condensers) which renders them not cleanable at all other than with circulating caustics; for such applications, successful long term operation may require special water filtration which adds cost.

Note: More often, long term operation will experience heat exchanger performance degradation from fouling as a result of a non-cleanable design.

The **fin-tube** heat exchanger is common in air heating and cooling coils (water-to-air) and for some heat recovery devices.

Other heat exchange methods exist that are optimized for the heat recovery media in use, such as heat wheels and drums, and heat pipes.

Building Operations and Maintenance 145

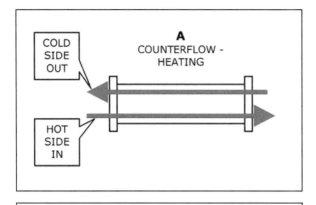

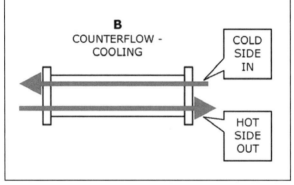

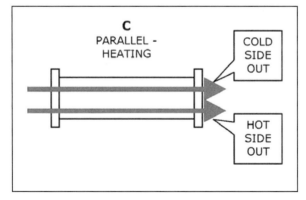

Figure 8-6. Heat Exchanger Approach Diagrams

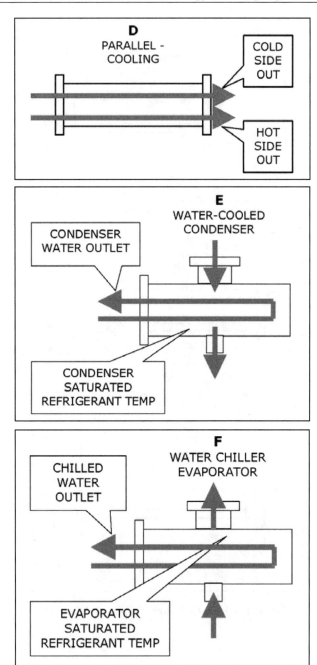

Figure 8-6. Heat Exchanger Approach Diagrams (*Continued*)

Building Operations and Maintenance 147

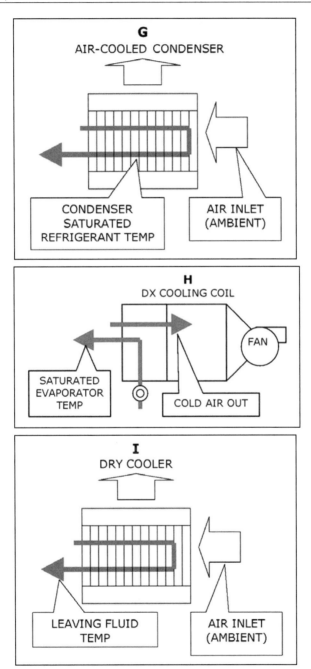

Figure 8-6. Heat Exchanger Approach Diagrams (*Continued*)

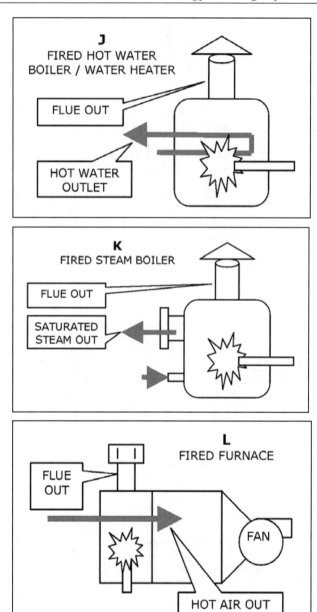

Figure 8-6. Heat Exchanger Approach Diagrams (*Continued*)

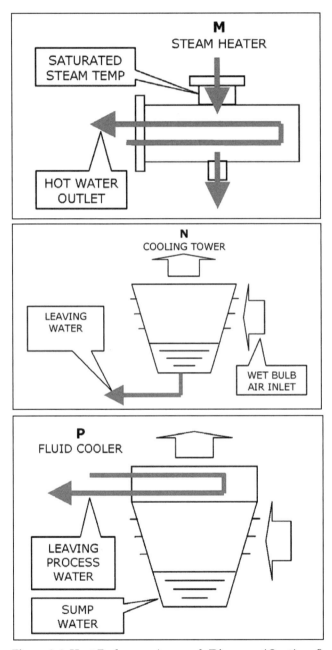

Figure 8-6. Heat Exchanger Approach Diagrams (*Continued*)

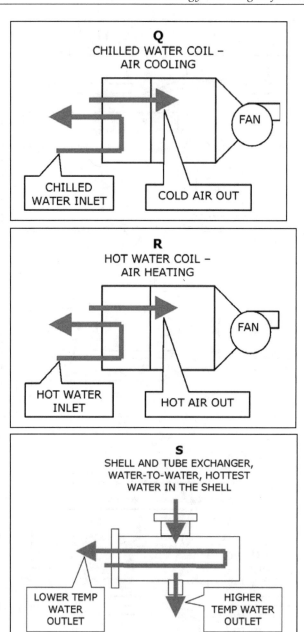

Figure 8-6. Heat Exchanger Approach Diagrams (*Continued*)

Building Operations and Maintenance 151

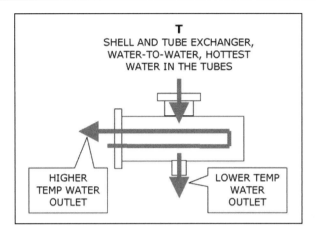

Figure 8-6. Heat Exchanger Approach Diagrams (*Continued*)

Chapter 9

Quantifying Savings

GENERAL

Business decisions to go or not go with energy improvement projects are normally based on simple payback or rate of return (ROI). Many other projects compete for available funds and so are ranked by need or by merit.

The success of transitioning from the report phase to actual constructive change depends largely on the ability to accurately predict cost and savings.

Computer modeling is commonly provided when guaranteed savings are used or when the dynamics involved make it impractical to estimate any other way. This method has the potential to be the most accurate and flexible for 'what if' scenarios, but is also time-intensive to implement.

In some cases, manual calculations or spreadsheets will do a satisfactory job, and examples of those are provided in this chapter.

In other cases, modeling or calculations are not effective, and these may rely on anecdotal rules of thumb or be ignored altogether. The more dynamic the process, the harder it is to estimate.

In all cases, understanding the principles at work is the foundation. For this reason, each individual ECM description begins with the statement of "Basis of Savings."

Of course, a barrier 'versus' all of these is cost, and determining the cost of the ECMs is beyond the scope of this text. The driver for most business decisions regarding capital improvements is cost vs. benefit. This section gives some examples of how to arrive at the first half of the equation.

Most calculations, including computer models, will include assumptions to simplify the work and consequently will have a plus/minus tolerance that should be understood to be there, regardless of the number of decimal places. On a good day, the calculations will yield results that are close to reality.

Understanding the processes at work sufficiently to apply correlations and equations in spreadsheets is necessary before attempting manual calculations. Established estimating modeling software has an advantage here because the research has been done in advance, leaving only application and parameter input to the user.

One source of equations for estimating savings of standard DSM measures is *Vol. 2: Fundamental Equations of Residential and Commercial End Uses*, EPRI, 1993.

Finally, there are some things that defy quantifying. How these items are treated is up to the reader. Some people ignore them entirely if they can't be quantified, while others rely on anecdotal rules of thumb to claim some savings.

This chapter will provide solved examples to illustrate some of the common concepts responsible for ECM savings.

ESTABLISHING THE HVAC LOAD PROFILE

This is priority #1 for estimating HVAC energy use. The more accurate the profile, the more accurate the predictions. To EXACTLY determine the profile requires a well calibrated computer model. It is possible to arrive at a reasonably close profile in many cases using a few assumptions. Sample **Figure 9-1** and **Figure 9-2** show a simplified model that assumes:

1. Cooling load is greatest at the highest outdoor temperature
2. Heating load is greatest at the lowest outdoor temperature
3. Cooling load will be zero at some value of outdoor temperature
4. Heating load will be zero at some value of outdoor temperature
5. Heating/Cooling loads will vary linearly between zero and maximum according to outside air temperature.

This simplified model works well when
- Thermal break-even temperatures are known (when the chiller is off, when the boiler is off)
- Cooling loads follow outside air dry bulb temperature

Step 1

Establish max heating and cooling loads. This can be done with load calculations or, if you are fortunate enough to be in the facility on the cold-

Quantifying Savings

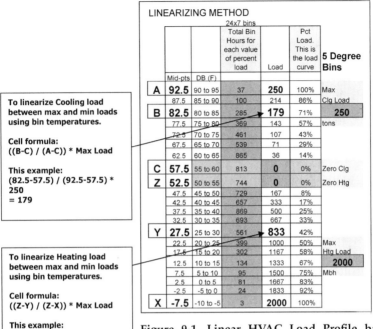

Figure 9-1. Linear HVAC Load Profile based on Outdoor Temperature—(55 deg F Break Even Temperature) Bins are for Colorado Springs. Note: Examples in this text use bins in five (5) degree increments to conserve space. It is more customary to use 2 degree interval bins for detailed calculations. Original spreadsheet shown below.

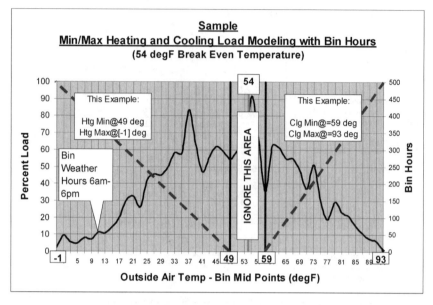

Figure 9-2. Pictorial Representation of Linear HVAC Load Profile
Dashed line shows assumed linear heating and cooling load variation.

est and hottest days, use observed actual values.

Use caution if attempting to calculate heating and cooling loads with a spreadsheet, especially cooling loads. If you do this, validate the output against actual loads. Unlike residential loads where skin loads (envelope) represent up to 70% of heating/cooing energy demand, commercial buildings have varying degrees of internal loads which complicate the estimation of cooling and heating load profile.

Step 2

Establish break-even temperature, which is where the internal heating is equal to the envelope loss; at outside temperatures above this cooling will be needed and at outside temperatures below this heating will be needed. This is not easy and depends upon thermal insulation, thermal mass, percent of glazing, and internal loads. One way to get a general idea is to review control settings and practices to see when boilers and chillers are turned on. Refer to **Chapter 11—Mechanical Systems**, Air-Side Economizer, **"Representative Break Even Temperatures for Different Building Activities."**

Step 3
Establish a deadband. Plus/minus (5) degrees F is suggested. There is uncertainty of whether heating or cooling (or neither or both) is going on in this region, so it is common practice to ignore any savings in this range of temperatures. In some buildings, both the heating and cooling operate in this range and so ignoring this area may not capture all of the energy use.

Step 4
Establish minimum heating and cooling temperatures. Minimum cooling (zero cooling) is assumed to be (5) degrees above the break even temperature. Minimum heating (zero heating) is assumed to be (5) degrees below the break even temperature.

Step 5
Between minimum and maximum cooling outdoor temperatures, the cooling load can be presumed to vary directly with the outside air temperature. This is a simplified model and ignores swings in humidity and internal loads.

Step 6
Between minimum and maximum heating outdoor temperatures, the heating load can be assumed to vary directly with outdoor temperature.

Step 7
Normalize the heating and cooling loads to 0-100%. This is the load curve, with coincident hours. See the sample spreadsheet in **Figure 9-1** and pictorial represetntation of the concept in **Figure 9-2.**

ADJUSTING THE HVAC LOAD PROFILE
FOR HUMID CLIMATES

Humid climates will require modifications to this simplified model:
- Adjust the load profile for cooling, if it is sufficiently humid to cause a persistent refrigeration at moderate temperatures.
- Add air reheating if a dehumidification cycle is used.

Humidity Modified Cooling Profile (Houston, TX)

	Mid-pts	DB (F)	Total Hrs	HR (gr/lb)	Load, from Temperature	Load, from Humidity	Required Cooling Load	Pct Load. This is the load curve
	97.5	95 to 100	18	112.7	250	250	250	100%
Dehum low range	92.5	90 to 95	307	116.8	222	250	250	100%
85	87.5	85 to 90	514	116.8	194	250	250	100%
50%	82.5	80 to 85	1114	110.7	167	202	202	81%
	77.5	75 to 80	1478	109	139	189	189	75%
Dehum high range	72.5	70 to 75	1432	96.7	111	92	111	44%
116.8	67.5	65 to 70	795	83.2	83	0	83	33%
100%	62.5	60 to 65	861	68.2	56	0	56	22%
	57.5	55 to 60	563	52.3	28	0	28	11%
	52.5	50 to 55	477	44.4	0	0		
Observed Maximum	47.5	45 to 50	369	36.4				
Cooling Load	42.5	40 to 45	281	31.6				
250 tons	37.5	35 to 40	305	27.5				
	32.5	30 to 35	162	22.3				
	27.5	25 to 30	57	16.7				
	22.5	20 to 25	16	12.4				
	17.5	15 to 20	10	7.9				
	12.5	10 to 15	1	6.7				

Criteria: When dry bulb is above 55 degrees, dehumidification may be needed.
for this example, 85 grains/lb requires 50% cooling and max grains/lb requires 100% cooling regardless of tempeature.

Figure 9-3. Modified HVAC Load Profile—Humid Climate

The cooling load profile can be modified to accommodate humid climates by including dew point temperature or absolute humidity (humidity ratio). Beginning at a nominal 75 deg F/50%rH space temperature, it can be reasoned that any outside air humidity that is higher than this will require some dehumidification. Using a nominal 50% refrigeration capacity as a minimum for dehumidification and full capacity needed for highest humidity levels, the demands for dehumidification can be calculated for all points in between. The, the overall cooling load profile will be the greater of the demand from temperature and dehumidification.

ADJUSTING THE HVAC LOAD PROFILE FOR OVERLAPPING HEATING AND COOLING

The load profile zone between heating and cooling is usually ignored due to uncertainty of what exactly is going on. Air economizers and temperature resets are active in this area, mild temperatures may have windows open and equipment off, etc. *If you are sure* there is consistent overlap (simultaneous heating and cooling), the load profile can be adjusted to show it. Linearizing from temperature in the overlap zone will overstate the amount of overlap, so a subjective factor for minimum persistent load is added to quantify the heat/cool loads even when weather does not dictate. **Figure 9-5** shows this method and the results are graphed in **Figure 9-4**.

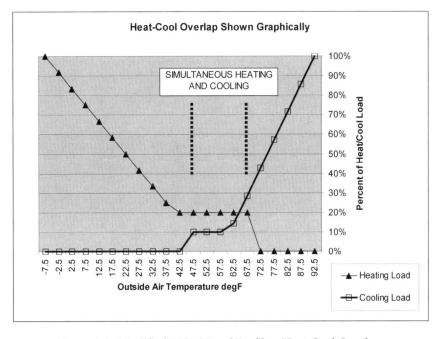

Figure 9-4. Modified HVAC Load Profile—Heat-Cool Overlap

OVERLAPPING HEATING AND COOLING EXAMPLE

		24x7 bins									
	Mid-pts	DB (F)	Total Bin Hours for each value of percent load	Cooling Load from Weather	Min Clg load during ovelap period	Clg Load	Pct Clg Load. This is the cooling load curve	Heating Load from Weather	Min Htg load during ovelap period	Htg Load	Pct Htg Load. This is the heating load curve
	92.5	90 to 95	37	250	25	250	100%	0	0	0	0%
Max Clg Load **250** tons	87.5	85 to 90	100	214	25	214	86%	0	0	0	0%
	82.5	80 to 85	285	179	25	179	71%	0	0	0	0%
	77.5	75 to 80	369	143	25	143	57%	0	0	0	0%
	72.5	70 to 75	461	107	25	107	43%	0	0	0	0%
Zero Htg	67.5	65 to 70	539	71	25	71	29%	0	400	400	20%
	62.5	60 to 65	865	36	25	36	14%	0	400	400	20%
Weather zero for Clg	57.5	55 to 60	813	0	25	25	10%	0	400	400	20%
Weather zero for Htg	52.5	50 to 55	744	0	25	25	10%	0	400	400	20%
	47.5	45 to 50	729	0	**0**	25	10%	167	400	400	20%
Zero Clg	42.5	40 to 45	657	0	0	0	0%	333	400	400	25%
	37.5	35 to 40	869	0	0	0	0%	500	400	500	33%
	32.5	30 to 35	693	0	0	0	0%	667	400	667	42%
	27.5	25 to 30	561	0	0	0	0%	833	400	833	50%
	22.5	20 to 25	399	0	0	0	0%	1000	400	1000	58%
Max Htg Load **2000** Mbh	17.5	15 to 20	302	0	0	0	0%	1167	400	1167	67%
	12.5	10 to 15	134	0	0	0	0%	1333	400	1333	75%
	7.5	5 to 10	95	0	0	0	0%	1500	400	1500	83%
	2.5	0 to 5	81	0	0	0	0%	1667	400	1667	92%
	-2.5	-5 to 0	24	0	0	0	0%	1833	400	1833	100%
	-7.5	-10 to -5	3	0	0	0	0%	**2000**	400	2000	

Min Clg Load **10%** in overlap zone

Min Htg Load **20%** in overlap zone

Figure 9-5. Modified HVAC Load Profile—Heat/Cool Overlap

Quantifying Savings

LOAD-FOLLOWING AIR AND WATER FLOWS VS. CONSTANT FLOW (VFD BENEFIT)

Basis of Savings: Reduced energy transport power (fan and pump power)

Assume the flow rate tracks the load for 50-100% load. Below 50% load it may not track directly since factors like minimum air flows and laminar coil conditions may begin to dominate.

This can be for any load following scenario. For illustration, HVAC load following will be used, i.e. as the heating or cooling load decreases, the air flow and/or water flow decreases proportionally.

Refer to **Figure 9-2**. The 100% cooling load condition represents the constant volume energy. The reduced percent load dashed line represents the fan or pump load-following potential improvement and energy reduction from using variable flow technology.

With load following, power consumption for any bin is the air [or water] Hp transport energy:

Fan horsepower (motor Hp):
HP(air)
$= CFM * TSP / (6356 * \text{fan eff} * \text{drive eff} * \text{motor eff})$
Where CFM = cubic feet per minute air flow and
TSP = total static pressure, in. w.c.

Pump horsepower (motor Hp):
HP(water)
$= GPM * HEAD / (3960 * \text{pump eff} * \text{drive eff} * \text{motor eff})$
Where GPM = gallons per minute water flow
HEAD = total resisting pressure, ft. w.c.

Refer to **Figure 9-6**. First calculate the water/air horsepower for the 100% case which is the constant volume baseline. Then in a parallel column calculate the water/air horsepower for the modulating case. It is suggested to limit the minimum load to 40%.

Reduction in flow for centrifugal devices are matched with reductions in rpm, within a reasonable range, such as down to 40%. Reductions in energy use generally follow the affinity laws, but a suggested conservative method is to modify the affinity law and use the square instead of the cube, as follows:

162   Commercial Energy Auditing Reference Handbook

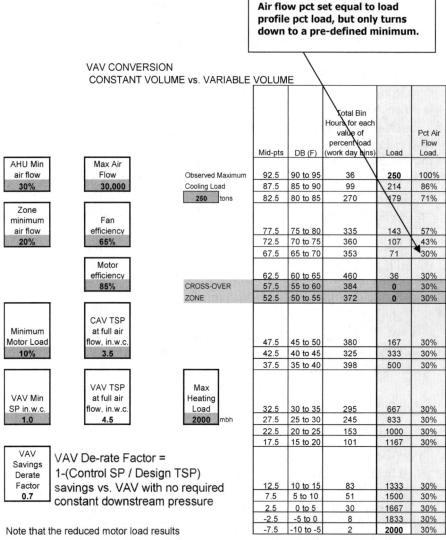

Figure 9-6. VAV Conversion

Quantifying Savings 163

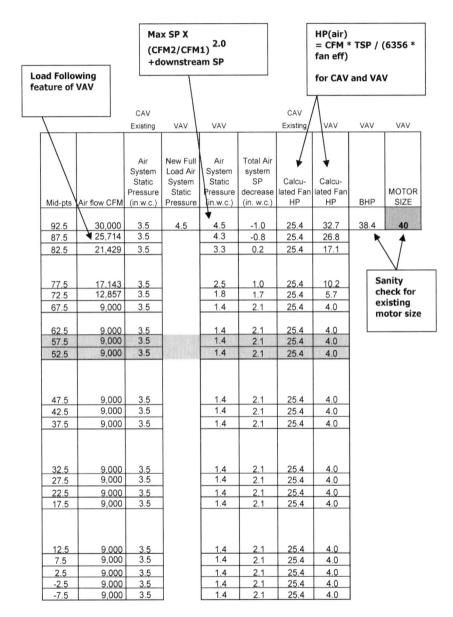

Figure 9-6. VAV Conversion (*Continued*)

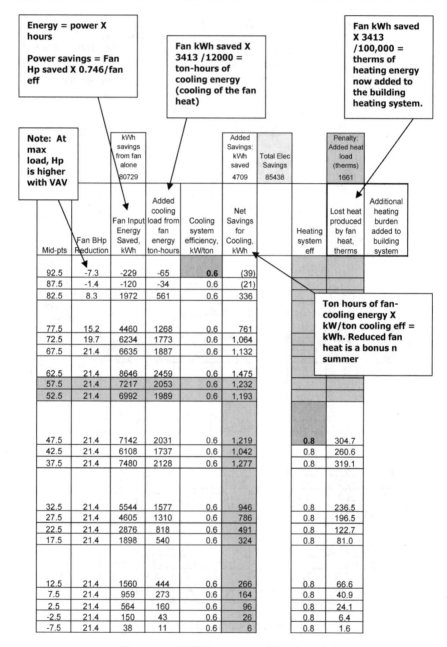

Figure 9-6. VAV Conversion (*Continued*)

Quantifying Savings

$$HP2 = HP1 \times (^{N2}/_{N1})^{2.0} \leftarrow \text{use square instead of cube}$$

For most commercial variable flow systems, there is a constant downstream pressure that is maintained, e.g. 1.0 in. w.c. for air or 50 ft. w.c. for water. In this case, the affinity law must be modified further to account for the downstream pressure and the work that requires.

See **Chapter 9—Quantifying Savings "VAV System Fan Savings Reduction for Maintaining Downstream Pressure."** This adjustment more accurately predicts the actual fan horsepower.

For each bin, there will be a horsepower (baseline and ECM) and a number of hours, and the remainder of the calculation is straight forward:

$$\text{Savings} = (\text{HP-hrs saved}) * \text{kW/Hp} * \$/\text{kWh}$$

VAV SYSTEM FAN SAVINGS REDUCTION FOR MAINTAINING DOWNSTREAM PRESSURE

For VAV air handling systems, there is a residual pressure requirement to operate the VAV boxes, and so the fan HP will not turn down fully according to the affinity laws. The de-rate factor for this depends on the ratio of the control point pressure being maintained (e.g. 1.0 in. w.c.) and the total design pressure TSP. The dynamic losses for ductwork, filters, coils are free to drop in pressure with the fan laws. The analysis method reduces system pressure by the fan laws but ADDS the residual pressure. The pressure reduction curve accounting for this constant downstream pressure requirement becomes

$$SP_2 = [SP_1 * (^{CFM2}/_{CFM1})^{2.0}] + CSP,$$

but not greater than the initial pressure.

Where CSP = the regulated constant downstream static pressure.

For VAV conversions with a required downstream constant pressure, the approximate de-rate is:

De-rate Factor = 1-(Control SP/Design TSP) = $1 - (^{CSP}/_{TSP})$
Where TSP = design total system static pressure.

HP savings for air/water flow reduction with de-rate factor shown.

$$HP_2 = [HP_1 * (^{N2}/_{N1})^{2.0}] * \underline{De\text{-}rate}$$

(This example is for a VFD so the exponent is 2.0 instead of 3.0. See also **Chapter 15—Fan and Pump Drives**, "Variable Speed Drive Considerations")

VAV Fan Savings De-rate Factors.
For the typical 1.0 in. w.c. control static pressure, the factors are as follows:

Control (CSP)	Total Design TSP	Factor $1 - (^{CSP}/_{TSP})$
1.0	3.0	0.67
1.0	4.0	0.75
1.0	5.0	0.80
1.0	6.0	0.83
1.0	7.0	0.86

SUPPLY AIR RESET VS. REHEAT—CONSTANT VOLUME

Basis of Savings:
1. Refrigeration system savings in cooling mode
   1-1.5 percent power reduction per deg F raised.

2. Reduced reheat energy penalty in heating mode.
   1.08 * altitude factor * CFM * dT
   where:
   CFM is the heating mode cfm
   dT is the difference between supply air temperature and room temperature (deg F)—i.e. the heating penalty incurred before any heating of the room begins.

Quantifying Savings 167

NOTES for constant volume systems:
- Air flow is constant
- Cooling load follows the cooling load profile
- Heating load follows the heating load profile, but includes additional reheat work from having to warm the supply temperature that is colder than room temperature.
- Heating energy savings dominate cooling savings by an order of magnitude and so the cooling energy savings can generally be neglected.
- Supply air reset in heating mode directly saves heating energy by reducing the reheat penalty.
- Whenever supply air is delivered below room temperature, and heating is required in the room, the heating load for that room is not the total heat required. The supply air must be heated as well.
- However, supply air reset only saves cooling energy if the source cooling unit is reset. Simply resetting the temperature of an air handler while using the same temperature of chilled water does not save cooling energy.
- Cooling below 55 degrees F outside air temperature is normally from air economizer and no refrigeration costs or supply air reset benefits occur.
- See Chapter 21—Formulas and Conversions, "Air Density Ratios (Altitude Correction Factors)"

SUPPLY AIR RESET WITH VAV VS. INCREASED FAN ENERGY

Basis of Savings:
1. Refrigeration system savings in cooling mode. (see NOTES for VAV Systems)
   1-1.5 percent power reduction per deg F raised.

2. Reduced reheat energy penalty in heating mode.
   *1.08 * altitude factor * CFM * dT*
   where:
   CFM is the heating mode cfm
   dT is the difference between supply air temperature and room temperature (deg F)—i.e. the VAV heating penalty incurred before any VAV reheat penalty of the room begins.

SUPPLY AIR RESET FOR CONSTANT VOLUME REHEAT

				Load modulates, but air flow is constant	
		Total Bin Hours for each value of percent		Heating and	Air Flow
Mid-pts	DB (F)	load	Load	Cooling pct load	(cfm)
92.5	90 to 95	37	**250**	100%	**50000**
87.5	85 to 90	100	214	86%	50000
82.5	80 to 85	285	153	61%	50000
77.5	75 to 80	369	87	35%	50000
72.5	70 to 75	461	37	15%	50000
67.5	65 to 70	539	11	4%	50000
62.5	60 to 65	865	2	1%	50000
57.5	55 to 60	813	0	0%	50000
52.5	50 to 55	744	0	0%	50000
47.5	45 to 50	729	92	8%	50000
42.5	40 to 45	657	183	17%	50000
37.5	35 to 40	869	275	25%	50000
32.5	30 to 35	693	367	33%	50000
27.5	25 to 30	561	458	42%	50000
22.5	20 to 25	399	550	50%	50000
17.5	15 to 20	302	642	58%	50000
12.5	10 to 15	134	733	67%	50000
7.5	5 to 10	95	825	75%	50000
2.5	0 to 5	81	917	83%	50000
-2.5	-5 to 0	24	1008	92%	50000
-7.5	-10 to -5	3	**1100**	100%	50000

CROSS OVER ZONE (rows 57.5 and 52.5)

Altitude Factor 1.0

Cooling set to cut out below 60 degF in this example
Boiler set to cut out above 50 degF in this example

Figure 9-7. Supply Air Reset for Constant Volume Reheat

Quantifying Savings

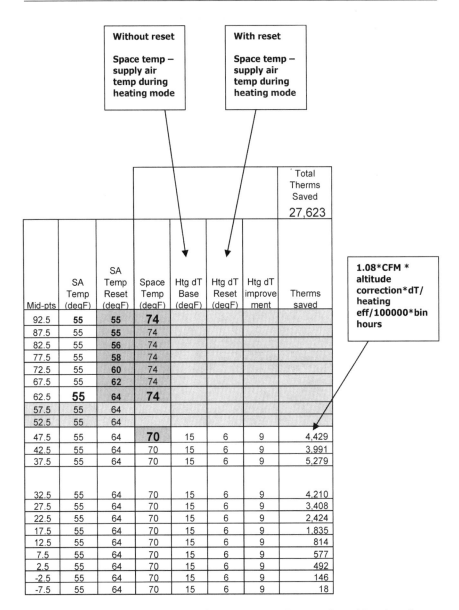

Figure 9-7. Supply Air Reset for Constant Volume Reheat (*Continued*)

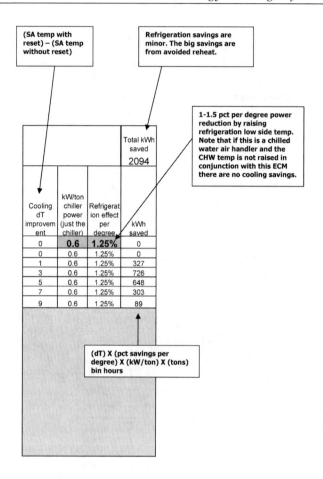

Figure 9-7. Supply Air Reset for Constant Volume Reheat (*Continued*)

Quantifying Savings 171

NOTES for Variable Air Volume (VAV) systems:
- Cooling load and air flow follows the cooling load profile
- Heating airflow is constant at a minimum setting (fraction) of the cooling cfm
- Heating load follows the heating load profile, but includes a penalty for heating the supply air temperature up to room temperature.
- Heating energy savings dominate, and cooling system savings are negligible. Cooling mode reset benefits are minor, due to the competing effects of increased fan energy.
- The higher the VAV box minimum air flows, the higher the savings.
- VAV reheat penalty. Whenever supply air is delivered below room temperature, and heating is required in the room, the heating load for that room is not the total heat required. The supply air must be heated as well. This is the VAV reheat penalty that is built into all single path VAV systems.
- Supply air reset in heating mode directly saves heating energy by reducing the reheat penalty.
- Supply air reset only saves cooling energy if the source cooling unit is reset. Simply resetting the temperature of an air handler while using the same temperature of chilled water does not save cooling energy.
- Cooling below 55 degrees F outside air temperature is normally from air economizer and no refrigeration costs or supply air reset benefits occur.

VAV cooling mode reset competing effects:
Air horsepower traded for refrigeration horsepower—VAV systems in cooling mode.

As supply air in cooling mode is reset upward, the chilled water temperature at the central plant can be reset upward and that creates savings from the refrigeration cycle:

For variable volume air systems, the higher supply air temperature results in a need for greater air flow for a given load and so fan horsepower increases and refrigeration work increases (from added fan work). For most reset schedules, especially those based on return air temperature the air penalties are usually higher than the refrigeration savings. Even for a very modest reset schedule based on outside air or percent cooling load, only a small portion of the apparent savings remains after the penalties are subtracted. **For this reason, unless M&V testing shows otherwise,** supply air reset for VAV systems should be limited to heating mode operation.

Air Flow Increase Factor:
The difference between space temperature and supply air temperature determines the amount of air required for a given load, so the factor for VAV air cfm increase is:

VAV Cooling Mode Reset Air Flow Increase Factor
$$= (RM - SA\ NO\ RESET / RM - SA\ RESET)$$
where:
RM = room temperature
$SA\ NO\ RESET$ = supply air temperature with no reset
$SA\ RESET$ = supply air temperature with reset

The additional air flow causes additional fan work
VAV cooling mode fan kW increase from reset
$$= 0.746 * HP1 \times (CFM2/CFM1)^{2.0}$$
where:
$HP1$ = fan hp with no reset
$CFM2$ = supply air flow with reset
$CFM1$ = supply air flow with no reset

The additional fan work ends up as heat in the air stream, increasing refrigeration load. (Any added energy transport loss—more fan heat—in cooling mode compounds itself by increasing refrigeration work.)

VAV cooling mode refrigeration load kW increase from fan kW increase
$$= kW\ increase \times (CH\ kW/ton)(3413/12{,}000)$$
where:
$kW\ increase$ = fan kW increase from reset
$CH\ kW/ton$ = chiller efficiency kW/ton

Quantifying Savings

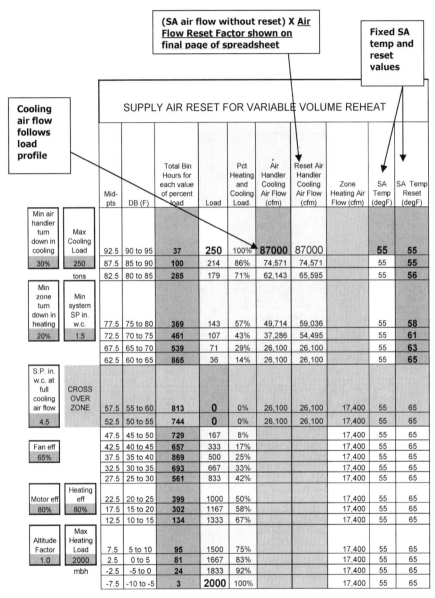

Figure 9-8. Supply Air Reset for VAV Reheat

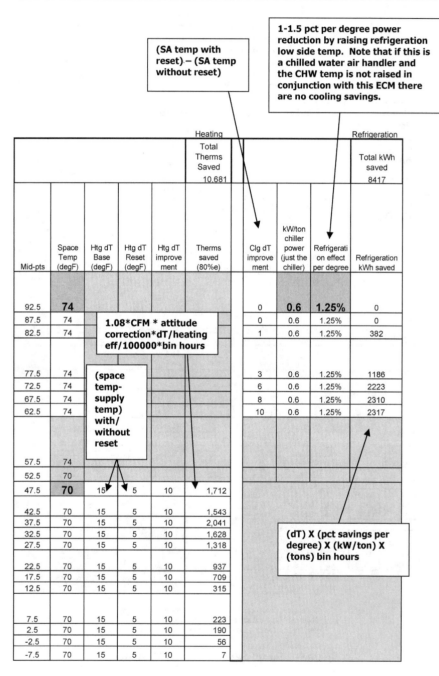

Figure 9-8. Supply Air Reset for VAV Reheat (*Continued*)

Quantifying Savings

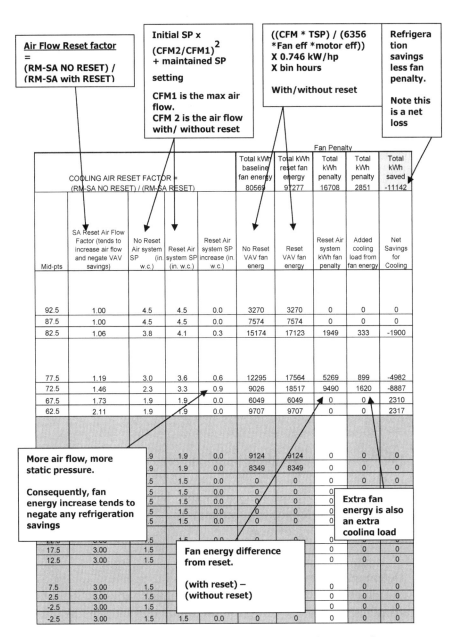

Figure 9-8. Supply Air Reset for VAV Reheat (*Continued*)

CONDENSER WATER RESET VS. CONSTANT TEMPERATURE

Basis of Savings: Reduced chiller power.
This is a balance between refrigeration power savings and added cooling tower fan power.
The key parameters are refrigeration load and coincident dry bulb and wet bulb temperature. Any apparent savings below 55 deg F dry bulb should be disregarded since the air economizer should be active and chillers locked out for most buildings. Within the occupied hours that are NOT lower than 55 deg F, identifying the refrigeration load and the coincident wet bulb temperature will support the necessary math to identify savings.

Step 1
Using a bin weather program, make the bins of occupied times, with dry bulb and coincident wet bulb temperatures.

Step 2
Block out all bin hours lower than 55 deg F dry bulb temperature.

Step 3
Apply zero cooling load to 55 degrees F and full design load (not nameplate capacity) to the highest bin temperature, and interpolate linearly between to establish the cooling load line.

Step 4
Determine the economical "approach" temperature to operate the cooling tower at. This is normally between 7-12 degrees, and depends upon the cooling tower fan energy per ton. This is a complex process but is made easier by the introduction of a term "Relative Cooling Tower Factor." An example is provided.

Step 5
Determine the lowest allowable condenser temperature for the chiller. This varies by manufacturer, but usually ranges from 60-70. Some chillers can operate on colder water, but most cannot.

Step 6
For each row of bin hours, determine the optimal condenser temperature from the relationship "wet bulb + approach, but not less than

Quantifying Savings

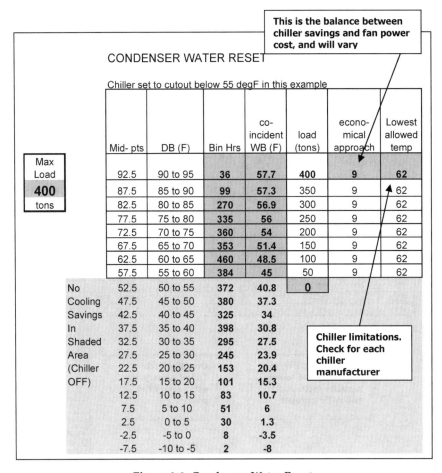

Figure 9-9. Condenser Water Reset

the equipment low limit." Compare the optimal setting to the base design fixed setting. For hours that the optimizing routine would have condenser temperatures lower than the base design, these are hours where savings will occur.

- For hours when the wet bulb temperature is much lower than the dry bulb temperature, savings for this measure are pronounced and approach the theoretical 1-1.5 percent per degree power reduction. But when the wet bulb depression is close to the economical approach value, the savings are reduced because power is added from cooling

Commercial Energy Auditing Reference Handbook

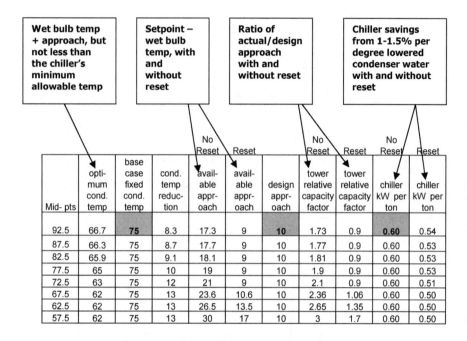

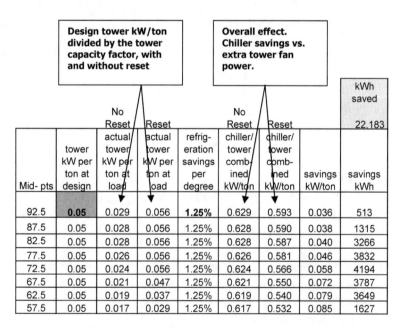

Figure 9-9. Condenser Water Reset (*Continued*)

tower horsepower to achieve the gains. Lower condenser water reduces chiller energy, but requires additional cooling tower fan horsepower. The lower the temperature, the higher the incremental cost of lowering the water temperature, so there is a balance point below which the overall energy consumption (chiller plus tower) no longer improves and may actually get worse. This dynamic can be approximated by using a "Relative Tower Capacity Factor."

The relative cooling tower capacity factor concept quantifies the fact that at drier conditions the cooling tower can achieve set point much easier, and that the condenser reset routine will work the cooling towers harder, spending more tower fan energy. Only if the gains in chiller power are greater than the expense of extra tower energy is this routine beneficial.

Relative Cooling Tower Capacity Factor
$= (\text{AVAILABLE APPROACH}/\text{DESIGN APPROACH})$
where:
AVAILABLE APPROACH = cooling tower set pt—actual wet bulb temperature
DESIGN APPROACH = design cooling tower set pt—design wet bulb temperature

And, cooling tower kW/ton at various conditions
= **Tower kW/ton @ design conditions/Cooling Tower Capacity Factor**

This will identify the cooling tower energy used. Combined with the chiller energy used, this provides the basis of evaluating savings.

Overall kW/ton for each bin of hours
=**Chiller kW/ton + cooling tower kW/ton**

- While the energy savings from condenser water reset near design conditions may be marginal, most of the chiller hours will be at wet bulb conditions that are more favorable, and this is where the energy savings are attained. The control strategy capitalizes on this by continuously adjusting the operating set point based on ambient wet bulb. For those chiller operating hours when the wet bulb is significantly lower than design and the cooling tower can easily produce colder water with little tower energy penalty, the savings will be much more pronounced and closer to the theoretical 1-1.5% per degree. Of course, the chiller load during these shoulder seasons or overnight periods are usually lower than maximum, so the chiller load profile and coincident wet bulb temperature is key to quantifying savings.

- Refer to the table in **Figure 9-10**. For a chiller with 0.5kW/ton efficiency, paired with a 0.07 kW/ton cooling tower, a proposed 5 degree reduction (from 12 to 7 degrees F approach), would yield *worse* energy consumption than leaving it at 12 degrees F, due to the high cooling tower fan energy penalty. This same example, with all things equal except a 0.04 kW/ton cooling tower, would save 0.35% per degree lower overall cooling energy. This example shows that while operating the cooling tower **at or near design wet bulb conditions, most or all of the theoretical chiller savings from condenser water reset will usually be negated by the added cooling tower energy use, unless the cooling tower is very efficient (0.04 kW/ton or less).**

Condenser Water Reset:
Effect of Cooling Tower Energy Use on Overall Energy Savings
- Remember, without the cooling tower fan energy expense, the savings from the chiller would be 1-1.5% per degree lowered. Note that without an efficient cooling tower, condenser water reset can make overall efficiency worse due to high tower fan penalty.

Cooling Tower kW/ton	Ratio of kW/ton Chiller to Tower (chiller 0.5 kW/ton)	Overall Cooling Energy Percent Savings per degF Lowered (chiller + cooling tower energy)	
0.03	16.7	0.63%	
0.04	12.5	0.35%	
0.05	10.0	0.08%	better
0.06	8.3	-0.18%	worse
0.07	7.1	-0.42%	
0.08	6.3	-0.66%	
0.09	5.6	-0.88%	
0.10	5.0	-1.10%	
Condenser Water Temperature Lowered from 12 to 7 deg F Approach			

Figure 9-10. Cooling Tower Fan Energy vs. Lower Condenser Water Temperature

- The "approach" value is a parameter of the optimized condenser temperature calculation, and should be selected based on the cooling tower in use, for best economy. Suggested values of cooling tower approach for best overall cooling efficiency (kW/ton) are shown in **Figure 9-11**. By using the "ratio" instead of specific combinations, this information can be applied to any combination of chiller and cooling tower. This is the value inserted in the sequence "…optimum cooling tower set point shall be equal to the calculated wet bulb temperature plus approach…," and provides further evidence of the importance of amply sized cooling towers with low fan Hp.

Chiller kW/ton	Tower kW/ton	Ratio of Chiller / Tower kW/ton	Typical Lowest Economical Tower Approach For Condenser Water Reset, degF
0.5	0.1	5:1	15.0
0.5	0.085	6:1	13.5
0.5	0.07	7:1	12.0
0.5	0.06	8:1	11.0
0.5	0.05	10:1	10.0
0.5	0.04	12.5:1	8.5
0.5	0.03	17:1	7.0

Figure 9-11. **Cooling Tower Sizing Effect on Optimum Condenser Water Temperature.** Amply sized cooling towers are required for achieving chiller savings without high cooling tower fan energy penalties.

CHILLED WATER RESET FOR VARIABLE PUMPING VS. INCREASED PUMP ENERGY

Basis of Savings: Reduced chiller power.

The savings are offset partly by added pump energy. As the load decreases, the water temperature is raised and the variable flow pumping compensates with increased flow.

Refrigeration system savings in cooling mode.
1-1.5 percent power reduction per deg F raised.

Water Flow Rate when load and dT are known

$$GPM = \text{Load}/(500 * dT)$$

Where GPM = gallons per minute water flow
Load = Btuh (design load Btuh * percent capacity)
dT = temperature difference between return and supply temperatures, assuming the return temperature stays the same. i.e. half the dT will require twice the water flow.

This acknowledges that flow reduction will track load downwards only to a point. In this case, 40% is assumed. Flow = (tons x 12,000) / (500 x dT)

Max flow = (450 x 12,000)/(500 x (55-45)) = 1080 gpm

Min flow = 40% x 1080 = 432

VARIABLE FLOW CHILLED WATER WITH TEMPERATURE RESET
Chiller set to cutout below 55 degF in this example

	Mid-Points	Range	Bin Hrs	Load (tons)	Pct Load	CHW temp	CHW temp	Return Temp	Minimum CHW flow for analysis
	92.5	90 to 95	36	**450**	100%	45	45.0	55	432
Max Load	87.5	85 to 90	99	394	88%	45	45.0	55	432
450	82.5	80 to 85	270	338	75%	45	46.0	55	432
tons	77.5	75 to 80	335	281	63%	45	47.0	55	432
	72.5	70 to 75	360	225	50%	45	48.0	55	432
Min. Linear	67.5	65 to 70	353	169	38%	45	49.0	55	432
Load	62.5	60 to 65	460	113	25%	45	50.0	55	432
40%	57.5	55 to 60	384	56	13%	45	51.0	55	432
No	52.5	50 to 55	372	**0**					
savings	47.5	45 to 50	380						
in the	42.5	40 to 45	325						
shaded	37.5	35 to 40	398						
area	32.5	30 to 35	295						
	27.5	25 to 30	245						
Chiller	22.5	20 to 25	153						
Off	17.5	15 to 20	101						
	12.5	10 to 15	83						
	7.5	5 to 10	51						
	2.5	0 to 5	30						
	-2.5	-5 to 0	8						
	-7.5	-10 to -5	2						

The reset schedule for this example is:

OA	CHWS
85	45
55	51

Figure 9-12. Variable Chilled Water Flow with Temperature Reset

Quantifying Savings

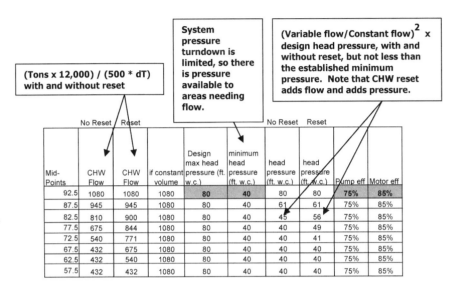

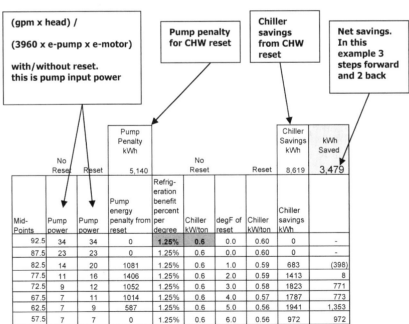

Figure 9-12. Variable Chilled Water Flow with Temperature Reset (*Continued*)

Pump horsepower (motor Hp):
$$HP(water) = GPM * HEAD/_{(3960 \, * \, pump \, eff \, * \, drive \, eff \, * \, motor \, eff)}$$
Where GPM, is gallons per minute water flow and HEAD is the total resisting pressure, ft. w.c.
What changes in this ECM is the GPM.
Assume the system head pressure remains the same.

The amount of pump penalty depends on the amount of reset, but losing about half of the benefit is typical. This is because as chilled water temperature is raised, additional flow is required for a given load.

$$GPM = (TONS * 12000/_{500 \, * \, dT})$$
where:
dT = differential temperature between return and supply, degrees F

The increase in chilled water flow from increasing supply temperature, with a fixed return temperature, is therefore directly proportional to the decrease in differential temperature.

Iterations with a spreadsheet can increase the ECM benefit by optimizing the two competing factors.

WATER-SIDE ECONOMIZER VS. CHILLER COOLING

Basis of Savings: Reduced chiller run time.
If air-side economizer is available, it should be used in lieu of water-side economizer for best energy efficiency. This is because the air-side uses only an air handler fan, while the water-side uses a cooling tower fan and condenser pumps and chilled water pumps—everything in the chiller plant other than the chiller itself. Water-side economizer is useful when no air-economizer exists or where the use of winter outside air is problematic, such as in museums and data centers.

The water-side economizer allows the chiller to be turned off and, instead, uses the cooling tower water indirectly through a heat exchanger. For this to work, the ambient wet bulb temperature must be sufficiently low to achieve the desired chilled water temperature—i.e. for 45 deg F chilled water a wet bulb temperature of 35 degrees or lower is usually required.

The following presumes air-side economizer is ruled out. Thus, cooling hours at all outside temperatures are viable.

Analysis compares cooling costs with a chiller to cooling costs with the water-side economizer.

The limitation of the flat plate is wet bulb temperature since that drives the cooling tower; however the A/C load on a building is influenced by dry bulb temperature. In most cases, the capacity of the flat plate in a building is greatest when the need for the free cooling is the least and so economic justification of these systems usually depends on there being some persistent winter cooling load that cannot be served by a standard air economizer.

Step 1
Identify the hours of operation the flat plate system could be used, the coincident load in tons, and the number of hours.

For envelope loads, use dry bulb bin weather data with coincident wet bulb temperatures, estimate the break even temperature, and interpolate between that point (zero cooling) and maximum cooling on the warmest bin temperature. Then rule out those bins that have coincident wet bulb temperatures too high to make the flat plate system work.

For high internal loads that exist all year long, such as a data center, the load is independent of weather and constant. This analysis only requires the number of hours of sufficiently low wet bulb that are concurrent with the cooling load.

Step 2
Base case, with the chiller. Identify the chiller and cooling tower kW; pump energy is the same in both cases and can be neglected. Note that for low ambient cooling tower operation, most cooling towers have 8-10% of their rated nominal capacity with no fan running, and at low wet bulb temperatures this effect will be amplified. Cooling tower energy use will be markedly different for base case and water-side economizer option.

Step 3
Alternate case, with water-side economizer. Chiller kW = 0, and cooling tower fan run time will be more. Identifying the capacity (in HVAC tons) of the cooling tower at low ambient wet bulb is key to this analysis. This can be approximated using the relative cooling tower capacity factor concept, which quantifies the fact that at drier conditions the cooling tower

can achieve set point much easier. During water-side economizer operation, the tower will be driving for a low water temperature with a low approach this will work the cooling towers harder, spending more tower fan energy.

Relative Cooling Tower Capacity Factor
= (AVAILABLE APPROACH/$_\text{DESIGN APPROACH}$)
where:
AVAILABLE APPROACH = cooling tower set pt—actual wet bulb temperature
DESIGN APPROACH = design cooling tower set pt—design wet bulb temperature

And, cooling tower kW/ton at various conditions
= Tower kW/ton @ design conditions/$_\text{Cooling Tower Capacity Factor}$

Using units of kW/ton for chiller and tower allows direct conversion from ton-hours to kWh.

Example: For a 35 degF wet bulb ambient, compare the work of the cooling tower for making 70 degree condenser water for a chiller vs. making 42 degree condenser water for the water-side economizer, ultimately to make chilled water for the building with the chiller off. The cooling tower is selected nominally for a 9 degree approach while making 70 degree leaving water and is rated at 0.08 kW/ton as paired to the chiller at design conditions.

Ans:
For the chiller, available approach is 70-35=35 degrees, and the factor is 35/9 = 3.8
Adjusted tower kW/ton is 0.08/3.8 = 0.02

For the water-side economizer, the available approach is 7, and the factor is 7/9 = 0.8
Adjusted tower kW/ton is 0.08/0.8 = 0.1 kW/ton

For each ton of cooling provided by the cooling tower in water-side economizer service at these conditions the tower will use five times as much energy than it would if serving the chiller at normal condenser temperatures.

This is still much less energy than running the chiller, but it does show the give and take of the water-side economizer. Note the advantage the simple air economizer has over the water-side economizer at most temperatures, since none of the chilled water system components are needed.

Quantifying Savings

Water –Side Economizer Analysis for Standard Building

The dry bulb bins establish the load profile in tons as usual. The coincident wet bulb bins establish the performance of the plate heat exchanger for each point in the load profile. This will tell when the heat exchanger can or can't provide the cooling.

WATER-SIDE ECONOMIZER SERVING BUILDING WITH MODERATE INTERNAL LOADS
This example assumes a 40 degree break even temperature, and a water-side economizer that will function at 37 degrees wet bulb

		Mid-pts	DB (F)	DB Bin Hrs	co-incident WB (F)	load (tons)	Chiller kW/ton	Nominal Tower kW/ton
Max Load 400 tons	No	92.5	90 to 95	36	57.7	400	0.6	0.08
	Savings	87.5	85 to 90	99	57.3	364		
	In Shaded	82.5	80 to 85	270	56.9	327		
	Area	77.5	75 to 80	335	56	291		
		72.5	70 to 75	360	54	255		
	Above 37 degF	67.5	65 to 70	353	51.4	218		
	Wet Bulb Tempera	62.5	60 to 65	460	48.5	182		
		57.5	55 to 60	384	45	145		
		52.5	50 to 55	372	40.8	109		
		47.5	45 to 50	380	37.3	73	0.6	0.08
		42.5	40 to 45	325	34	36	0.6	0.08
	No	37.5	35 to 40	398	30.8	0		
	Savings	32.5	30 to 35	295	27.5			
	In Shaded	27.5	25 to 30	245	23.9			
	Area	22.5	20 to 25	153	20.4			
		17.5	15 to 20	101	15.3			
	Below	12.5	10 to 15	83	10.7			
	Building	7.5	5 to 10	51	6			
	Break Even	2.5	0 to 5	30	1.3			
	Temperature	-2.5	-5 to 0	8	-3.5			
		-7.5	-10 to -5	2	-8			

NOTE there are good savings per hour but very few hours per year that are viable for this example. Without a very dry climate or a very low break even temperature, this system has limited value in serving ordinary building loads

Figure 9-13. Water-Side Economizer with Weather-Dependent Load

188 *Commercial Energy Auditing Reference Handbook*

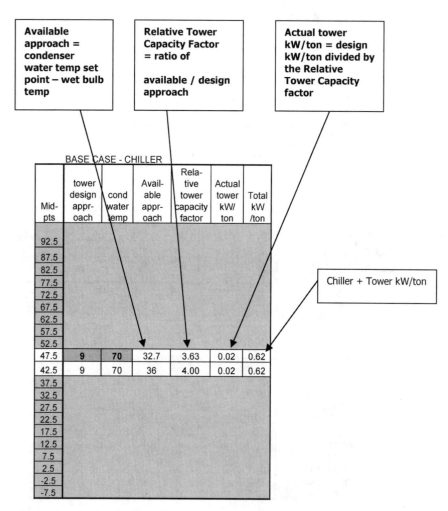

Figure 9-13. Water-side Economizer with Weather-dependent Load (*Continued*)

Quantifying Savings

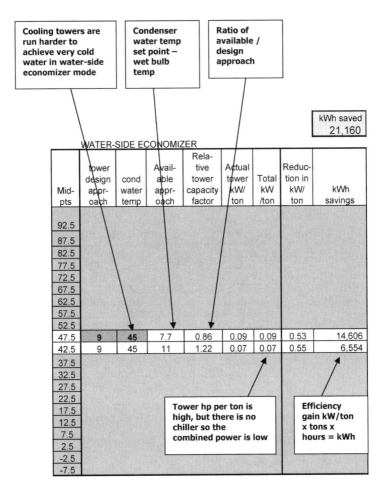

Figure 9-13. Water-side Economizer with Weather-dependent Load (*Continued*)

Water-side Economizer Analysis for Data Center, 24x7 internal load, nearly weather independent (see Figure 9-14).

Water-Side Economizer Analysis for Data Center, 24x7 internal load, nearly weather independent.

WATER-SIDE ECONOMIZER SERVING BUILDING WITH HIGH AND WEATHER INDEPENDENT INTERNAL LOADS
This example assumes NO break even temperature, and a water-side economizer that will function at 37 degrees wet bulb

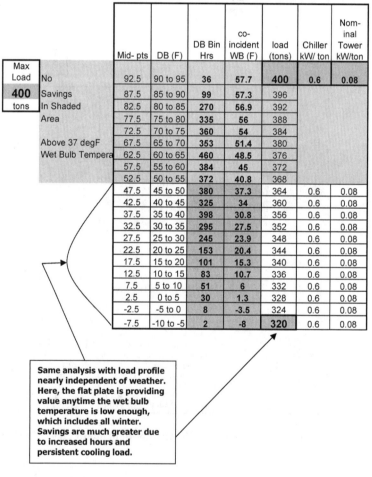

		Mid- pts	DB (F)	DB Bin Hrs	co-incident WB (F)	load (tons)	Chiller kW/ ton	Nominal Tower kW/ton
Max Load 400 tons	No Savings In Shaded Area	92.5	90 to 95	36	57.7	400	0.6	0.08
		87.5	85 to 90	99	57.3	396		
		82.5	80 to 85	270	56.9	392		
		77.5	75 to 80	335	56	388		
		72.5	70 to 75	360	54	384		
	Above 37 degF Wet Bulb Tempera	67.5	65 to 70	353	51.4	380		
		62.5	60 to 65	460	48.5	376		
		57.5	55 to 60	384	45	372		
		52.5	50 to 55	372	40.8	368		
		47.5	45 to 50	380	37.3	364	0.6	0.08
		42.5	40 to 45	325	34	360	0.6	0.08
		37.5	35 to 40	398	30.8	356	0.6	0.08
		32.5	30 to 35	295	27.5	352	0.6	0.08
		27.5	25 to 30	245	23.9	348	0.6	0.08
		22.5	20 to 25	153	20.4	344	0.6	0.08
		17.5	15 to 20	101	15.3	340	0.6	0.08
		12.5	10 to 15	83	10.7	336	0.6	0.08
		7.5	5 to 10	51	6	332	0.6	0.08
		2.5	0 to 5	30	1.3	328	0.6	0.08
		-2.5	-5 to 0	8	-3.5	324	0.6	0.08
		-7.5	-10 to -5	2	-8	320	0.6	0.08

Same analysis with load profile nearly independent of weather. Here, the flat plate is providing value anytime the wet bulb temperature is low enough, which includes all winter. Savings are much greater due to increased hours and persistent cooling load.

Figure 9-14. Water-side Economizer with Weather-independent Load

Quantifying Savings 191

	BASE CASE - CHILLER					
Mid-pts	tower design approach	cond water temp	Available approach	Relative tower capacity factor	Actual tower kW/ton	Total kW/ton
92.5						
87.5						
82.5						
77.5						
72.5						
67.5						
62.5						
57.5						
52.5						
47.5	9	70	70	7.78	0.01	0.61
42.5	9	70	70	7.78	0.01	0.61
37.5	9	70	70	7.78	0.01	0.61
32.5	9	70	70	7.78	0.01	0.61
27.5	9	70	70	7.78	0.01	0.61
22.5	9	70	70	7.78	0.01	0.61
17.5	9	70	70	7.78	0.01	0.61
12.5	9	70	70	7.78	0.01	0.61
7.5	9	70	70	7.78	0.01	0.61
2.5	9	70	70	7.78	0.01	0.61
-2.5	9	70	70	7.78	0.01	0.61
-7.5	9	70	70	7.78	0.01	0.61

Figure 9-14. Water-side Economizer with Weather-independent Load (*Continued*)

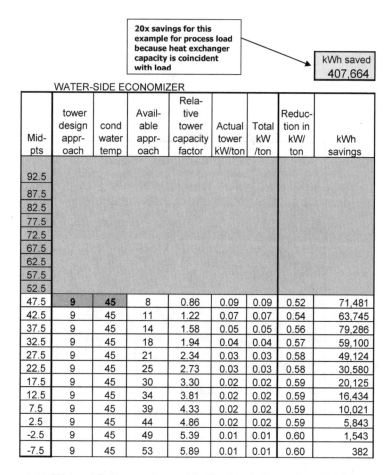

Figure 9-14. Water-side Economizer with Weather-independent Load (*Cont'd*)

Quantifying Savings 193

HIGHER EFFICIENCY LIGHTING VS. EXISTING LIGHTING

Basis of Savings: Incremental source efficiency improvement

Hours of operation * differential kW of lighting = kWh savings.

Break this down by area and don't over-estimate the hours of operation. (8760 hours per year is usually unrealistic). See **Chapter 16—Lighting "Average Lighting Hours by Building Type"** for typical hours of operation for lighting in different sectors.

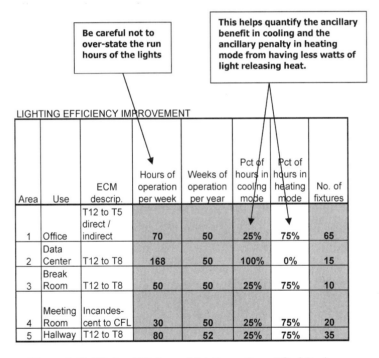

Figure 9-15. Higher Efficiency Lighting—Quantified Savings

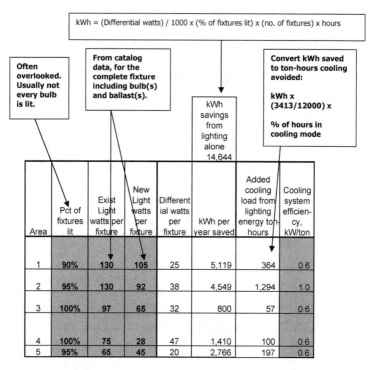

Figure 9-15. Higher Efficiency Lighting—Quantified Savings (*Continued*)

Quantifying Savings

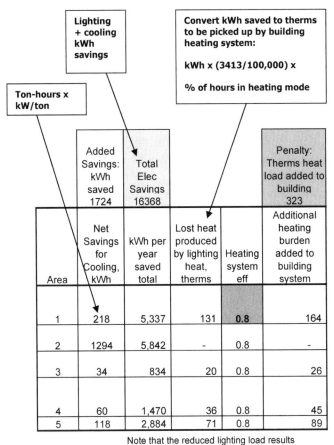

Figure 9-15. Higher Efficiency Lighting—Quantified Savings (*Continued*)

HIGHER EFFICIENCY MOTORS VS. EXISTING MOTORS

Basis of Savings: Incremental source efficiency improvement
Load profile is required to properly analyze this and to not overestimate savings.

Equipment efficiency changes at different loads can usually be neglected since both the existing and competing motor will have a similar pattern.

For a motor that is load following for HVAC, the same method of estimating pct load with bin weather can be used. For motors that are constant load, the energy impact is simply the motor size * percent load. Note that motors are seldom operated at full nameplate load, so the "percent load" factor applies to all motor calculations. Using nameplate motor hp alone will almost always over-state energy use and savings.

A reasonable estimate of percent motor load can be obtained with a simple amp reading, noting the ratio of measured amps to full load nameplate amps.

$$\text{Approx Motor pct load} = {}^{\text{(Measured Amps)}}/_{\text{(Full Load nameplate Amps)}}$$

$$\text{Motor kW input} = {}^{\text{(HP) * (\%Load) * (0.746)}}/_{\text{motor eff}}$$

NOTE: Motors, like all machines, have internal losses. These are minor compared to full load output, but become more pronounced at very low loads. **A motor with nothing connected to the shaft will not consume zero work.** For scenarios that produce very low estimated percentage of load (turndown), a factor for "minimum motor load" is appropriate.

Evaluation method when motor load profile and hours are known (see Figure 9-16).

Sample HVAC Motor Evaluation Method using Bin Hours for Motor Load Profile (see Figure 9-17).

Quantifying Savings

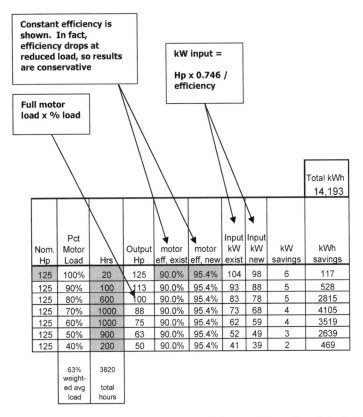

Figure 9-16. Higher Efficiency Motor with known Motor Load Profile

Evaluation method when motor load profile and hours are known.

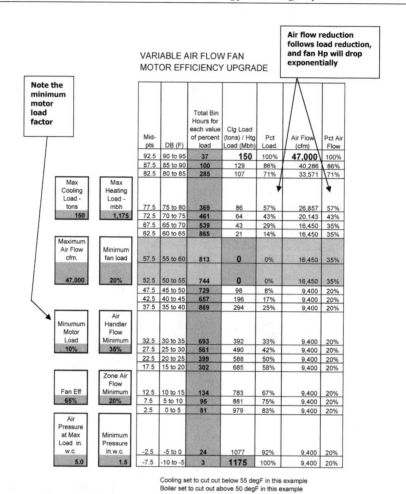

Figure 9-17. Higher Efficiency Motor With Bin Weather Load Profile

Sample HVAC motor evaluation method using bin hours for load profile.

Quantifying Savings

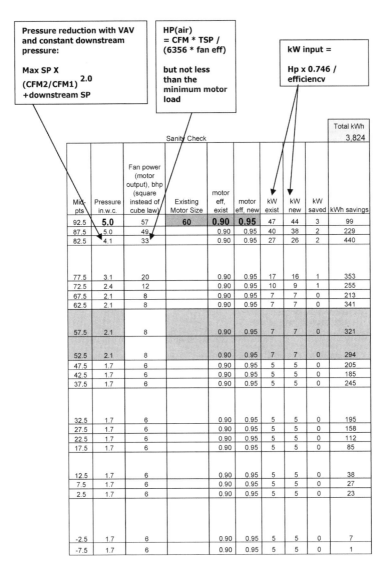

Figure 9-17. Higher Efficiency Motor With Bin Weather Load Profile (*Cont'd*)

HIGHER EFFICIENCY CHILLER VS. EXISTING CHILLER

Basis of Savings: Incremental source efficiency improvement
Load profile is required for proper analysis.

Ton-hours * kW/ton = kWh

For comfort cooling chillers, weather bins are a reasonable way to estimate the load profile. Establish maximum load at the highest bin, and then estimate the break even point below which the chiller will be off.

For process chillers (other than HVAC cooling), load profile will be determined by other than weather.

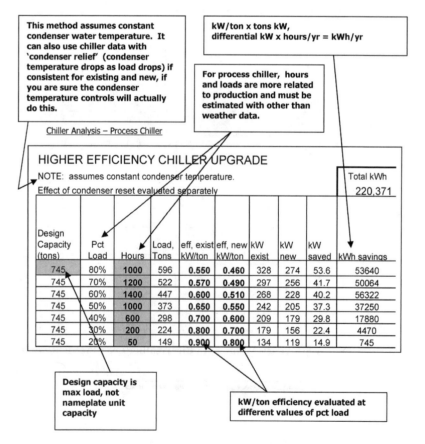

Figure 9-18. Higher Efficiency Chiller—Process Chiller

Quantifying Savings

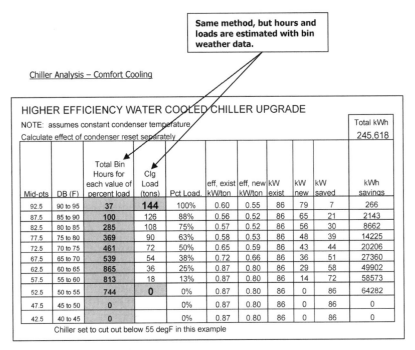

Figure 9-19. Higher Efficiency Chiller—Comfort Cooling

HIGHER EFFICIENCY BOILER VS. EXISTING BOILER

Basis of Savings: Incremental source efficiency improvement

Load profile is required to properly analyze this and to not overestimate savings.

For comfort heating boilers, weather bins are a reasonable way to estimate the load profile. Establish maximum load at the lowest bin, and then estimate the break even point above which the boiler will be off.

For process boilers (other than HVAC heating), load profile will be determined by other than weather.

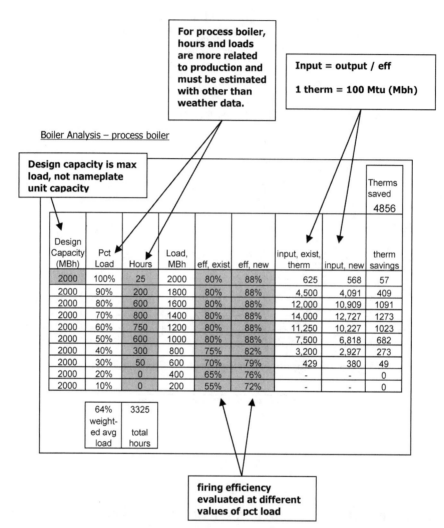

Figure 9-20. Higher Efficiency Boiler—Process Boiler

Quantifying Savings

Boiler Analysis – Comfort Heating

Same method, but hours and loads are estimated with bin weather data.

HIGH EFFICIENCY BOILER UPGRADE

mbh

Mid-pts	DB (F)	Total Bin Hours for each value of percent load	Load Profile	Pct Load	eff, exist	eff, new	input, exist, therm	input, new	Total Therm Saved therm saving
									18,0!
62.5	60 to 65	865	0	0%	0.50	0.58	-	-	0
57.5	55 to 60	813	357	7%	0.50	0.58	5,807	5,006	801
52.5	50 to 55	744	714	14%	0.50	0.58	10,629	9,163	1466
47.5	45 to 50	729	1071	21%	0.50	0.58	15,621	13,467	2155
42.5	40 to 45	657	1429	29%	0.50	0.58	18,771	16,182	2589
37.5	35 to 40	869	1786	36%	0.60	0.68	25,863	22,820	3043
32.5	30 to 35	693	2143	43%	0.70	0.78	21,214	19,038	2176
27.5	25 to 30	561	2500	50%	0.70	0.78	20,036	17,981	2055
22.5	20 to 25	399	2857	57%	0.80	0.88	14,250	12,955	1295
17.5	15 to 20	302	3214	64%	0.81	0.89	11,984	10,907	1077
12.5	10 to 15	134	3571	71%	0.82	0.90	5,836	5,317	519
7.5	5 to 10	95	3929	79%	0.82	0.90	4,551	4,147	405
2.5	0 to 5	81	4286	86%	0.83	0.91	4,182	3,815	368
-2.5	-5 to 0	24	4643	93%	0.81	0.89	1,376	1,252	124
-7.5	-10 to -5	3	5000	100%	0.80	0.88	188	170	17
-12.5	-10 to -15	0		0%	0.50	0.58	-	-	0
-17.5	-15 to -20	0		0%	0.50	0.58	-	-	0

Boiler set to cut out above 60 degF in this example

Figure 9-21. Higher Efficiency Boiler—Comfort Heating

HOT WATER RESET FROM OUTSIDE AIR VS. CONSTANT TEMPERATURE

Basis of Savings: *Reduced standby losses (thermal loss through piping)*

Assume a constant indoor temperature to simplify the analysis. Standard tables of insulated pipe heat loss can be used for different temperatures. These estimates will usually be conservative, since seldom are heating water piping systems 100% insulated and bare pipe and fitting losses are much higher.

Heat loss during winter is not a complete loss. The heat isn't in the place you normally will want it, but if it is inside the insulated envelope at least part of it is beneficial.

The method shown provides an approximation of savings, but has several subjective parameters that should be acknowledged, such as the beneficial heat factor.

Step 1

Determine the number of hours of operation for the associated outside air temperatures.

Step 2

Estimate the length of pipe involved. 2 inch pipe is shown in this example, but other sizes would be considered as well in practice.

Step 3

Estimate the fraction of the piping system that is insulated. Assuming it is 100% insulated is usually not realistic.

Step 4

Determine the desired reset schedule.

Step 5

Determine the heat loss for bare pipes and insulated pipes at fixed and reset temperatures.

Quantifying Savings

Bin hours are grouped into 10-degree increments, for convenience of calculating piping heat losses.

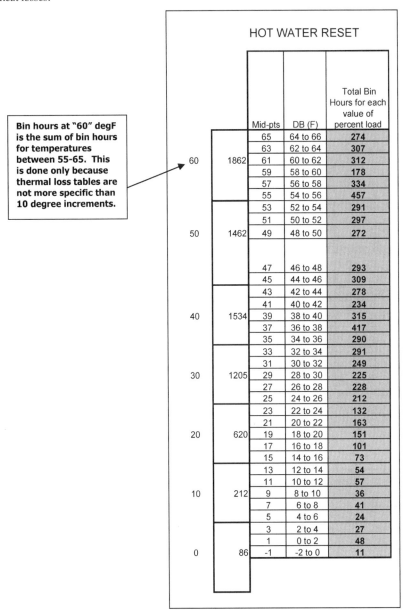

Figure 9-22. Hot Water Reset from Outside Air

206 Commercial Energy Auditing Reference Handbook

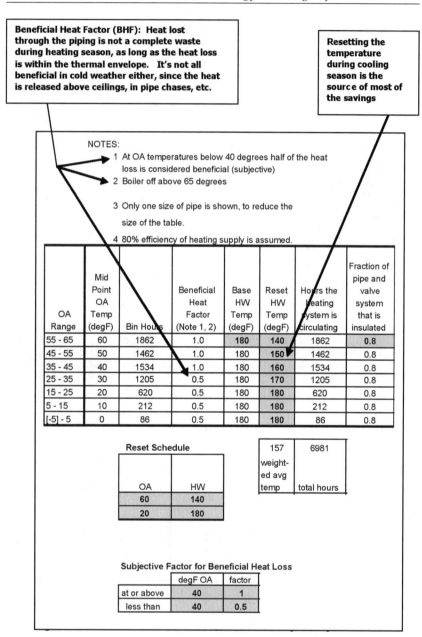

Figure 9-22. Hot Water Reset from Outside Air

Quantifying Savings

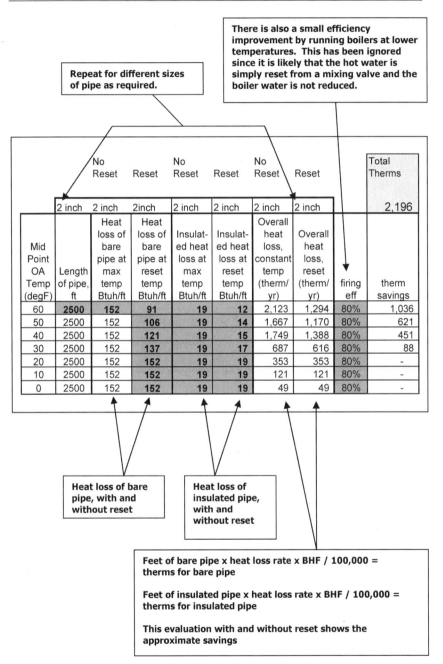

Figure 9-22. Hot Water Reset from Outside Air

REDUCE AIR SYSTEM FRICTION LOSSES—CONSTANT VOLUME

Basis of Savings: Reduced fan horsepower.
Since airflow is constant, Hp is constant. What is needed are the total hours and maximum air flow.
The same calculation applies to constant flow water systems.

Note: This same calculation applies to variable volume systems, with the flow profile added. With variable air/water flow, Hp is variable roughly as the square of the flow reduction. Savings will depend on the number of hours with high flow rates (e.g. cooling mode), since air horsepower is only significant during these times. For climates with a low number of cooling hours, savings will be minimal compared to cities with a large number of cooling hours.

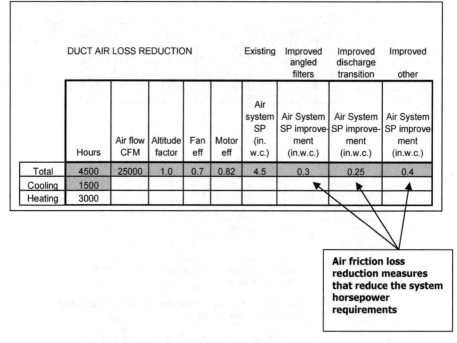

Figure 9-23. Reducing Air System Friction Losses—Constant Volume

Quantifying Savings

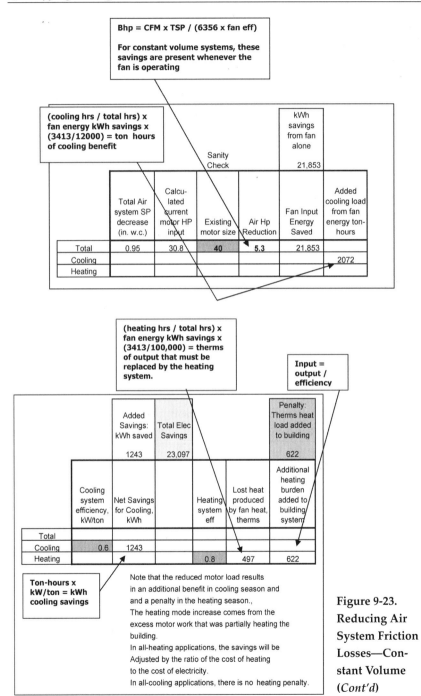

Figure 9-23. Reducing Air System Friction Losses—Constant Volume (Cont'd)

COMPUTER MODELING

General

There are a variety of computer software programs to use. Some are proprietary, some are public domain, some are produced by equipment manufacturer, including:
- DOE-2 and various overlays such as Visual DOE and eQuest
- Trane Trace
- Carrier HAP
- Market Manager
- BLAST
- Energy Plus

Calibrating and Testing the Computer Model

Using the simulation for a year gone by, compare:
- Total annual use matches actual utility bills
- Seasonal rise and fall of heating/cooling energy matches actual usage pattern

Sanity Checks

- Actual vs. modeled maximum Loads.

    If the model produces maximum heating and cooling loads, compare these to actual equipment sizes for a sanity check. Note that many HVAC systems include a measure of redundancy in the primary equipment, so establishing the 'actual' max heating and cooling load is needed. For example, asking the operators if both boilers run during the coldest days, and if so does the lag boiler run hard or cycle. Actually being on site during the hottest or coldest days is an excellent way to establish max loads, by noting the percent capacity of a modulating machine or percent run time for an on-off machine. For example, if the boiler runs 20 minutes and is off 10 minutes, it is running at 2/3 capacity and the actual load at that point is 2/3 of the boiler's rated capacity.

- Check numbers for heating and cooling systems.

    Heating and cooling capacities in terms of "Btuh per SF" (heating) and "SF per ton" (cooling) are good sanity checks, although these numbers—especially heating—vary by locale.

    The more important number for energy calculations is the annual en-

Quantifying Savings

	COMMERCIAL
Cooling Max Load Units: SF/Ton This is cooling system output, not labeled equipment size.	Source: ASHRAE Pocket Guide, 2001, Lo-Avg-Hi. © American Society of Heating, Refrigerating and Air-Conditioning Engineers, Inc., www.ashrae.org. **Figures in ( ) are Denver area estimates** Apartments, High Rise **450-400-350 (600)** Auditoriums, Churches, Theatres. **400-250-90 (500)** Schools, Colleges, Universities **240-185-150 (350)** Hotels, Motels, Dormitories **350-300-220 (500)** Libraries and Museums **340-280-200 (450)** Office Buildings (general) **360-280-190 (550) (380 for all glass walls)** Restaurants **135-100-80 (200)** Department stores **350-245-150 (400)** Malls **365-230-160 (475)**
Heating Max Load Units: kBtu/SF This is heating system output, with redundant equipment factored out. Input is output divided by efficiency.	Source: Author experience **30** (post 70's) (range 15-45) **45** (pre-70's) (range 22-67)
Air Flow Density Units: CFM/SF	Source: Author experience **0.8** range 0.5-2.25

Figure 9-24. Check Numbers for Cooling and Heating Design Loads

ergy use. For heating this is "Btu per year," and for cooling this is "ton-hours per year." Publishing check numbers for annual heating and cooling energy use, especially cooling, are risky since there are a great number of variables, including vintage, locale, user settings, hours of operation, equipment condition, outside air, and internal loads.

Heating loads are easier to quantify than cooling since most buildings with natural gas use most of it for heating.

MEASUREMENT AND VERIFICATION (M&V)

General

A full treatment of this topic is beyond the scope of this book. There are many excellent texts on the subject, most notably the IPMV 2001 [Volume 1], *International Performance Measurement & Verification Protocol* (IPMV), http://www.ipmvp.org.

When reviewing these, bear in mind that much of it is written to serve the performance contracting industry where guaranteed savings contracts are used and it is essential to define mutually agreeable rules of engagement and minimize disagreement on the outcome, which usually manifests itself in the form of 'who owes who' some money. The purpose of M&V for performance contracting is to provide a basis for fair business practice, especially on the part of the lending party who is assuming most of the risk.

The concept of M&V is very useful to energy savings projects. M&V serves to
- Validate the project's worth and confirm initial estimates.
- Verify and improve the forecasting skills of the energy professional

Note that M&V costs add to project costs, or more accurately subtract from project savings. For this reason, M&V methods and rigor should be balanced with actual need, and these parasitic costs minimized to the extent possible. A good rule of thumb is that M&V costs should never exceed 5% of the project cost, and 1% is more reasonable for most projects.

Measurement and verification can easily be part of a new control system, and can provide conclusive data for project performance including the controls themselves. EMS M&V will be of little use in justifying the project in advance, since the data provided is, by definition, after the fact. Still, collecting the information is extremely valuable and strongly encouraged, since the demonstrated savings can provide the impetus for additional improvements. Adding detailed notes of before and after conditions of the test, and other assumptions, will add credibility for the results.

Quantifying Savings

IPMVP 2001 M&V Options—Summary

Source: "Maximizing Energy Savings with Energy Management Systems," *Strategic Planning for Energy and the Environment*, Winter 2005.

M&V OPTION	DESCRIPTION
Option A: Partially Measured Retrofit Isolation	Savings are determined by partial field measurement of energy use by the system to which an energy conservation measure (ECM) was applied. Partial measurement refers to the fact that some (but not all) key parameters may be stipulated rather than measured, assuming the total impact of possible errors will not significantly affect the resulting savings. Measurements may be short-term or continuous.
Option B: Retrofit Isolation	Savings are determined by comprehensive field measurement of energy use by the system to which an ECM was applied. Measurements may be short-term or continuous and are taken throughout the post-retrofit period.
Option C: Whole Facility	Savings are determined by taking energy measurements at the whole facility level. Measurements may be short-term or continuous and are taken throughout the post-retrofit period.
Option D: Calibrated Simulation	Savings are determined through simulation of the energy use of component systems or the whole facility. Energy use simulation is calibrated with utility billing data and / or end-use metering.

Figure 9-25. IPMVP 2001 M&V Options—Summary

Establishing the M&V Baseline

This simple word is a source of great consternation in the world of M&V, since it is what subsequent measurements are compared to and actual savings are calculated from. This is especially true in performance contracting since the calculated differential is the basis of money transactions. If we had the luxury of being able to replay a period of time, changing only one thing (the project scope), then establishing the baseline would be easy; but this is not the case. In practice, we **measure what it is after** the fact, but often only **estimate what it would have been before**, and it is this estimating process that is the source of the consternation. Since there are any number of approaches and assumptions possible, and none of them perfect, what normally prevails in performance contracting is an agreement of what the baseline is assumed to be.

BASELINE METHOD	APPLICATION
Measure equipment full load efficiency before, and after, and then assume an annual load profile to calculate before and after usage	Large, identifiable points of energy use, such as chillers, boilers, large banks of lighting, etc. Note that this method disregards ancillary equipment and part-load performance.
Average building use for 10 days prior, and compare that average to the 'test day.' Possibly eliminate wild-card days such as Mondays, Fridays, pre/post-holidays, other special considerations.	Demand limiting
Normalize building energy usage for the test period against the same period one year prior, adjusting usage based on occupancy data and weather data	Building heating and cooling modifications where weather and occupancy have a strong influence on energy use
Normalize manufacturing energy use for the test period against a different period of same duration, adjusting usage based on manufacturing product throughput and days of production.	Processes dependent on manufactured product volume, and independent of weather
Computer modeling of alternate building designs, with annual energy estimating, to show differentials, either in magnitude or proportion.	This is often done to determine preferred system designs

Figure 9-26. Establishing a Reasonable Baseline—Examples

For performance contracting, this 'baseline' negotiation is done in advance of the work, so that calculating the resulting performance is an arithmetic process without ambiguity.

In the absence of a performance contracting environment, the contract issues are gone but the stakes are still high and establishing a proper baseline is still a key issue.

Chapter 10

Sustaining Savings

TENDENCY FOR INITIAL SAVINGS TO DETERIORATE

There is a significant body of evidence that initial energy savings from conservation projects tend to drift downward, eroding 10-20% of the initial savings within the first two years.
Source
1. Toole and Claridge, "Review on Persistence of Commissioning Benefits in New and Existing Buildings," International Conference for Enhanced Building Operations—Maximizing Building Energy Efficiency and Comfort, 2006.
2. "Is Commissioning Once Enough?," David Claridge et al, *Energy Engineering*, 2004.

Project Risk
Long-term payback assumptions are at risk from savings that are not persistent. It also represents risk for facilities that get funding for projects and subsequently get their operating budgets reduced by an amount equal to the initial savings. An awareness of this tendency for eroding savings is important, so that steps can be taken to mitigate it and sustain the benefits.

MAINTAINING INITIAL SAVINGS

Some ways to help prevent this backsliding are:
- Conservative calculations, however this merely 'hides' the savings.
- Measures that are immune to this; e.g. building shading elements.
- User buy-in and acceptance; measures that aren't too complicated for the people who will inherit them.
- Designs with ample provisions to encourage maintenance, especially the cleanability of heat transfer surfaces. This includes equipment that is easy to dismantle for cleaning, ready access and ample room for the work. Fouling factors in equipment selections add heat transfer surface area to the equipment and extend the service intervals.
- Clear and ample documentation so the project intent is not lost.

- Repeating training, testing, measurements and adjustments (Re-Commissioning).
- Management support is a key ingredient for success for any ECM related to a behavioral change. This facility director said it well: "To get energy standards to stick you will need support at the very highest level of the organization otherwise the doers have the responsibility without the authority and are doomed to failure."

CHECKLIST FOR SERVICE ACCESS AND OPERATIONS

The following was taken from a mechanical designer quality control (QC) checklist that serves to stress, from the beginning, the importance of service access for project success and sustained performance.

- User input received during the design development process, for buy-in?

- Drawings show service clearances shown with dotted lines to claim space?
  — filter and motor removal zones
  — coil pulls
  — tube pulls
  — compressor removal
  — automatic control panel access
  — above-ceiling item access
  — control valve access, etc.
  — motor controls, VFDs, etc.

- Can all equipment be reasonably and safely accessed for normal servicing ? Including:
  — general inspection
  — measurement and verification
  — testing and balancing
  — filter replacement
  — belt adjustment and replacement
  — motor replacement
  — lubrication
  — coil cleaning
  — fan cleaning

Sustaining Savings 217

- — control panel and control device access
  — valve access
  — pump seal repair
  — nozzle cleaning, etc.

- Other access considerations:
  — Access doors shown in rigid walls and ceilings for access, and are they large enough to work through?
  — Access doors upstream and downstream of each duct coil, duct humidifier, etc. for inspection and cleaning?
  — Can equipment located over a lay-in ceiling be accessed from the floor by a step ladder?
  — Permanent access to the roof for roof mounted equipment, preferably from indoors.
  — All equipment, coils, and control valves have unions or flanges for removal.
  — Control valves and dampers with position indication visible from floor.
  — Thermometers, pressure gages, and indicating instruments visible from the floor.
  — Shutoff valves and flanges/unions provided in a location to allow coil or equipment removal (e.g. not within the path of the coil pull).
  — Shutoff valves provided to isolate all equipment.
  — Drain, fill and vent provisions provided at all equipment.
  — Drain, fill and vent provisions provided for piping systems.
  — For piping that is racked (above ceilings, in chases, in tunnels, etc.) be sure each individual pipe system can be serviced or repaired. It should not be trapped by other pipes.
  — Blank space between successive air handler coils sections for cleaning.
  — Removable sections of insulation provided for items requiring routine servicing, such as pumps, strainers, control valves, etc.
  — Service platforms provided for equipment on steep roofs.
  — Service platforms provided for large or suspended equipment.

- Other owner-friendly design considerations:
  — Equipment drawn to scale? Will equipment fit in the rooms as shown?

- Can equipment be removed from the building after the building is built ?
- For equipment with multiple acceptable sources of supply, have the clearances been designed to accommodate each of the substitutions ?
- Equipment specified to be clearly labeled (tagged) to match the drawings.
- Outside air intakes with insect screens that are removable and cleanable.
- Water meters for system leak indication for hydronic make-up.
- Hour meters for system leak indication for centrifugal chiller purge unit, fire protection dry pipe air compressor, etc.
- Are schematic drawings included for each air and water system ?
- Are schematic drawings included for each fire alarm, security, and other specialty electrical system?

- Control system acceptance and survivability features:
  - Identify the level of sophistication the operators will accept.
  - Make provisions for the system to be user-friendly.
  - Accommodate owner preferences for materials, system supplier, man-machine interface, etc.
  - Identify skill level and training needs.
  - Gages and thermometers by key instruments.
  - Test wells adjacent to pipe and duct sensors for calibration.

- Automatic control system user-friendly interface provisions:
  - Control drawings located by each panel.
  - Clear and detailed sequences of operation.
  - Means to view set point, input, and output for each control loop.
  - Means to shut off panel power/air supply.
  - Devices labeled to match the control drawing.
  - Means to temporarily override outputs for maintenance purposes.
  - DDC: Display-adjust feature or plug-in for portable interface.
  - Conventional: Gages/dials to read temperatures, pressures, etc.
  - Pneumatic: output gages marked to reflect device position.

Section II
General Information

Chapter 11

Mechanical Systems

GENERAL

In commercial buildings, on average, 25% of total energy used is HVAC. Of that, 20%-60% is fan/pump transport energy.
Source: E Source, 2004
"Right sizing" is important since over-sizing moves the operating point of percent load for the equipment to a lower value that may cause inherent inefficiency.

RELATIVE EFFICIENCY OF AIR CONDITIONING SYSTEMS

Source: *Energy Management Handbook*, 6th Ed, Chap 10, Turner/Doty

GLOSSARY OF BASIC HVAC SYSTEM TYPES

Source: "Energy Conservation with Comfort," Honeywell, except where noted. "Courtesy of Honeywell International Inc."

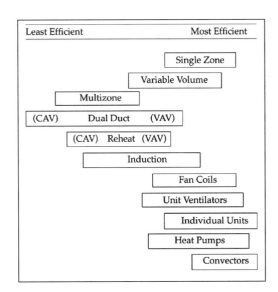

Figure 11-1. Relative energy efficiency of air-conditioning systems. Source: *Energy Management Handbook*, 6th Ed, Chap 10, Turner/Doty

221

SYSTEM	DESCRIPTION
Single Zone System Constant Volume.	Single zone systems consist of a mixing, conditioning and fan section. The conditioning section may have heating, cooling, humidifying or a combination of capabilities. Single zone systems can be factory assembled rooftop units, or built up from individual components and may or may not have distributing duct work.
Terminal Reheat System Constant Volume.	Reheat systems are modifications of single zone systems. Fixed cold temperature air is supplied by the central conditioning system and reheated in the terminal units when the space cooling load is less than maximum. The reheat is controlled by thermostats located in each conditioned space.
Multi-zone System Constant Volume.	Multi-zone systems condition all air at the central system and mix heated and cooled air at the unit to satisfy various zone loads as sensed by zone thermostats. These systems may be packaged roof-top units or field-fabricated systems.
Dual Duct System Constant Volume or Variable Volume.	Dual duct systems are similar to multizone systems except heated and cooled air is ducted to the conditioned spaces and mixed as required in terminal mixing boxes.
Variable Volume System	A variable air volume system delivers a varying amount of air as required by the conditioned spaces. The volume control may be by fan inlet (vortex) damper, discharge damper or fan speed control. Terminal sections may be single duct variable volume units with or without reheat, controlled by space thermostats.
Induction System Constant Volume.	Induction systems generally have units at the outside perimeter of conditioned spaces. Conditioned primary air is supplied to the units where it passes through nozzles or jets and by induction draws room air through the induction unit coil. Room temperature control is accomplished by modulating water flow through the unit coil.
Fan Coil Unit Constant Volume.	A fan coil unit consists of a cabinet with heating and/or cooling coil, motor and fan and a filter. The unit may be floor or ceiling mounted and uses 100% return air to condition a space.
Unit Ventilator Constant Volume.	A unit ventilator consists of a cabinet with heating and/or cooling coil, motor and fan, a filter and return air-outside air mixing section. The unit may be floor or ceiling mounted and uses return and outside air as required by the space.

Figure 11-2. Glossary of Basic HVAC System Types

Source: "Energy Conservation with Comfort," Honeywell, except where noted. "Courtesy of Honeywell International Inc."

Mechanical Systems

SYSTEM	DESCRIPTION
Unit Heater Constant Volume.	Unit heaters have a fan and heating coil which may be electric, hot water, or steam. They do not have distribution duct work but generally use adjustable air distribution vanes. Unit heaters may be mounted overhead for heating open areas or enclosed in cabinets for heating corridors and vestibules.
Perimeter Radiation Natural Convection	Perimeter radiation consists of electric resistance heaters or hot water radiators usually within an enclosure but without a fan. They are generally used around the conditioned perimeter of a building in conjunction with other interior systems to overcome heat losses through walls and windows.
Hot Water Converters	A hot water converter is a heat exchanger that uses steam or hot wat4er to raise the temperature of eating system water. Converters consist of a shell and tubes with the water to be heated circulated through the tubes and the heating steam or hot water circulated in the shell around the tubes.
	These system descriptions are not from the referenced Honeywell publication.
Heat Pump	A system that recovers heat using a refrigeration cycle. Control valves in the heat pump refrigerant circuit are analogous to a package window unit that can be turned around in different seasons depending upon where the heat is needed to be added or extracted. Heat source/sink may be water, air, or earth. Heat pumps can deliver heated or cooled water or air. The term "Heat Pump Balance Point" applies to air-source heat pumps only. The outside air temperature at which the increasing heat load of the building served is just equal to the decreasing capacity of the heat pump unit. Below this value of outside air temperature, supplemental resistance heat is required.
Single Zone VAV	Low pressure cooling only system that acts like "one big VAV box" This system normally serves a single large space with constant temperature variable volume air flow in response to space temperature.
Dedicated Outside Air System (DOAS)	Outside air is treated and ducted separately from the main recirculating air system. The energy advantage to this method, when combined with a VAV system, is the ability to set VAV box minimums to zero and eliminate the built-in simultaneous heating / cooling penalty.
Ground Source Heat Pump (GSHP)	An array of water-source heat pumps serving individual zones are connected by a common circulating water system that is buried in the earth. The earth becomes both a heat sink (summer, heat rejection) and a heat source (winter, heat absorption). Deep burial of the loop piping creates efficiency in winter since the ground temperature at these depths is unaffected by surface weather temperatures. Care must be taken to evaluate the cyclic adding and subtracting of heat in the loop field so that long term creep of soil temperatures that surround the loop piping can be predicted and planned for. Some systems that have dominant cooling loads (rejecting heat to the earth) will rise in temperature over time and may warrant an auxiliary cooling tower to equalize the seasonal heat flux. The long term drift of soil temperature will affect loop water temperature and thus the efficiency of the heat pumps, and failing to recognize this can result in over-stated energy benefits over the system life.

Figure 11-2. Glossary of Basic HVAC System Types (*Cont'd*)

SYSTEM	DESCRIPTION
Water Source Heat Pump Loop	An array of heat pumps serving individual zones are connected by a common circulating water system. During cold weather, the loop receives heat from an auxiliary heating source (boiler). During hot weather, the loop receives cooling from an auxiliary cooling source (cooling tower). The efficiency benefit of this system occurs during mild weather when some units call for heat and some call for cooling. In this range of temperatures, the heat from one zone is moved to another via the loop. To leverage this advantage, the loop temperature controls require a wide dead band to keep the auxiliary heating and cooling equipment off as much as possible.

Figure 11-2. Glossary of Basic HVAC System Types (*Cont'd*)

WATER-COOLED VS. AIR COOLED

Air-cooled equipment may be chosen for simplicity and lowest first cost, but from an energy standpoint air-cooled heat rejection is inherently inefficient compared to water-cooling. An air-cooled chiller with 70 degree F entering air may still operate at 110 degrees condensing temperature, a 40 degree approach, compared to water cooling with a 5-degree approach or less.

While water cooling has a fundamental advantage for heat transfer and efficiency, there are costs and operational considerations that come with water cooling. In some areas, water may be prohibitively expensive or be of such low quality that equipment damage will result. It is not uncommon for the cost of water/sewer to erode half or more of the energy savings.

From a sustainability view point, the water used for evaporation is recycled through rain, although from a practical standpoint the return point for the water may not be convenient. Water is not consumed through evaporative cooling like a fossil fuel but may have considerations for local supply and infrastructure.

THERMAL ENERGY TRANSPORT NOTES

- Rule of thumb for HVAC energy transport is that you can pump energy from point A to point B at about one fourth of the energy cost to blow it. Source: *Energy Engineering*, Vol. 102, No. 4, 2005. The basis for this is the fact that the specific heat (C_p) of water (1.0) is about four times that of air (0.24), so it takes four times as many pounds of

Mechanical Systems 225

air to carry the energy.
- The energy downfall of many district heating systems is the distribution piping thermal losses during reduced loads. These may account for 10%-15% of full load heating burden, but may account for 50% of total heating burden when in mild weather.
- Managing transport energy as a fraction of total heating and cooling energy is suggested. These include HVAC air systems, and central or district heating/cooling plants. Keeping the cost of transporting the thermal energy in proper proportion with the thermal energy itself is an obvious goal. Energy "budget" guidelines for matching these two are shown in the **Appendix** item " **Facility Guide Specifications: Suggestions to Build-In Energy Efficiency**."

Note: Once constructed, the opportunities to improve on this are limited. For example, if pipes or ducts are downsized to save initial cost, this is very hard to justify removing after the fact and are seldom changed.

- High distribution energy inputs (either fans or pumps) is a larger concern in cooling mode than heating mode, since almost all of the energy put into the circulator ends up in the water or air and adds directly to the cooling load, and these become parasitic losses. In heating systems, the pump/fan energy is beneficial heat.

CHILLERS

Chilled water systems are popular for a number of reasons:

- Water can transport a unit of cooling (or heating) more compactly in piping than with ducts.

- Suitable for remote location of central equipment, along with associated maintenance and noise.

- Cooling efficiency much higher than package rooftop equipment when water-cooled centrifugals are used.

- Scalable and flexible, to accommodate building additions and changes.

Process	Compressor type	Compressor Sub-Type	Size range	Air-Cooled (95 degF)	Water-Cooled
Vapor compression	Positive displacement	Reciprocating	5-100	1.2 - 1.4 (COP 3.0-2.5)	0.8-0.9 (COP 4.5-4.0)
Vapor compression	Positive displacement	Screw / Scroll	50-400	1.2 - 1.4 (COP 3.0-2.5)	0.65-0.8 (COP 5.4-4.5)
Vapor compression	Kinetic	Centrifugal	200-4000	1.2 - 1.4 (COP 3.0-2.5)	0.4-0.6 (COP 8.9-5.9)
Chemical absorption cycle.	N/A	N/A	200-2000	N/A	**COP (0.7-1.3)

Figure 11-3. Typical Chiller Efficiencies
Units are kW/ton (COP)
**Absorption units are rated in terms of COP from applied heat input. kW/ton rating does not apply to absorbers.

Centrifugal Chiller Notes
- Centrifugal chillers are popular due to compact size, high COP (high efficiency), robust nature, high reliability, and user friendly reputation.

- COP at 0.5 kW/ton   (1 ton x 12,000 Btu)/(0.5 kW x 3412 Btu/kW)
     = 7.0

- Efficiencies fall off rapidly below 50%, unless condenser water reset is used, so staging control to keep operating load above 50% is very important.

- Chillers that can accept very cold condenser water (55 deg F) and have cooling needs coincident with reduced wet bulb temperatures such as in the Southwestern US, when it is easy to achieve colder condenser water temperatures, can achieve 10-15% annual overall cooling energy savings using condenser water reset, compared to a fixed temperature set point of 70 degrees F.

Absorption Chiller Notes
- Often very attractive if used with waste heat.

Mechanical Systems 227

- Existing absorption chillers that use a 'new energy' heat source (not waste heat) are usually good candidates for removal, in favor of an electric chiller.

- Barriers to use are high first cost, large footprint, operational problems at part loads (best use is base load cooling), and high annual maintenance. Usually impractical if 'new' energy is used for the heat source (compare COP to mechanical refrigeration). Single effect units can run on low grade heat, and 15 psi steam. Double effect units require higher grade heat, or 100 psi steam.

- Single Effect:
  Steam Use   18-20 lbs per hour per ton
  COP, single effect (1 ton x 12,000 Btu)/18 lb x 1000 Btu per lb
                      = 0.67

- Double effect:
  Steam Use   9-10 lbs per hour per ton
  COP, double effect (1 ton x 12,000 Btu)/9 lb x 1000 Btu per lb
                      = 1.33

- Cooling tower water use with absorption cooling: 33% more cooling water required, 4 gpm per ton is typical.

Auxiliaries
- Water-cooled chillers have auxiliary equipment that is required to make it work, such as cooling towers, condenser pumps and chilled water pumps. It is interesting to note that the efficiency rating of a chiller does not include the cooling tower fan energy or condenser pump energy. By contrast, air-cooled HVAC cooling equipment does include the energy spent in the condenser fans. Good practice will look at the entire chilled water system and not just the compressor, and will keep a proportional balance of the two.

Rule of Thumb for Auxiliaries
- The chiller auxiliaries should each have an energy budget of not more than 1/10th of the power requirement of the compressor. For example, if the chiller is 500 kW, the tower fan, condenser pump, and all chilled water pumping combined should not each use no more

than 50 kW. Or, using cooling efficiencies and an example of 0.5 kW/ton:

0.5 kW/ton	Compressor
0.05 kW/ton	Chilled Water Pumping (primary and secondary combined)
0.05 kW/ton	Condenser Pumping
0.05 kW/ton	Cooling Tower Fan

PART LOAD CHILLED WATER SYSTEM PERFORMANCE

- Chillers are rated at both full load and part load. The standard part load test is IPLV.

IPLV

- IPLV stands for "Integrated Part Load Value." Since chiller equipment seldom runs at full load and efficiency varies with load, a system of part-load evaluation was developed by ARI. The basis is the weighted average weather data of 29 cities across the U.S. The assumed run times at different loads for standard IPLV ratings are:

100% load	1%
75% load	42%
50% load	45%
25% load	12%

Source: Air-Conditioning and Refrigeration Institute (ARI) 550/590-2003

- The energy downfall of many central chilled water systems is part load performance. It is typical that the auxiliaries (pumps and cooling tower fan) make up 20-30% of the total power requirement at full load. Unless provisions are made for the auxiliary device input to throttle with the cooling load, the proportions will shift at lower loads so the auxiliaries form a much larger constituent energy user. The cooling tower is normally cycled on a thermostat or has a VFD so its energy use should track the chiller load profile. But where pumping is "constant volume" the pump energy will be constant even when the chiller load is low.

Mechanical Systems

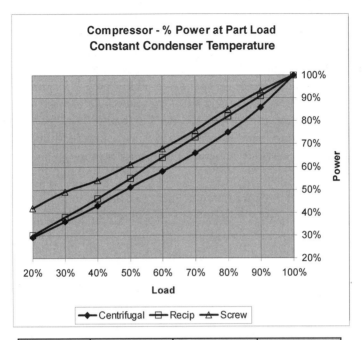

pct load	Centrifugal	Recip	Screw
100%	100%	100%	100%
90%	86%	91%	93%
80%	75%	82%	85%
70%	66%	73%	76%
60%	58%	64%	68%
50%	51%	55%	61%
40%	43%	46%	54%
30%	36%	38%	49%
20%	29%	30%	42%

Figure 11-4. Refrigeration Compressor Percent Power at Part Load
Source of data for reciprocating and screw compressors: P.C. Koelet, "Industrial Refrigeration—Principles, Design and Applications," 1992, reproduced with permission of Palgrave Macmillan.

Part Load Benefits of Variable Flow Pumping—Chillers

At full load, auxiliary components of a chilled water system are usually an order of magnitude less kW per ton of output cooling than the chillers. However, at part load cooling, constant flow pumps contribute a higher and higher percentage of the overall kW power consumption, and lower the overall cooling efficiency of the chilled water system.

By converting constant flow control to load-following variable flow, considerable improvements will be had during part load operation. For best economy, this will be standard practice for new designs. For retrofits,

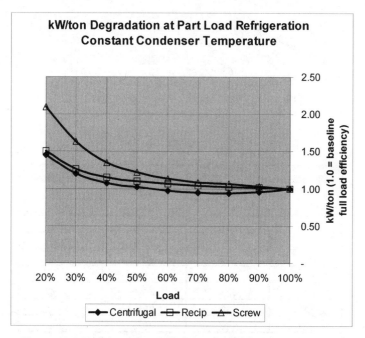

Figure 11-5. Refrigeration Efficiency Degradation at Part Load
Source of data for reciprocating and screw compressors: P.C. Koelet, "Industrial Refrigeration—Principles, Design and Applications," 1992, reproduced with permission of Palgrave Macmillan.

Mechanical Systems

there are operational considerations and this is an engineered change. For chilled water, flow control valves, air entrainment, and chiller tube velocity are considerations. For condenser water, control is usually from chiller condenser leaving temperature. In both cases, the excess flow keeps chiller heat transfer at its peak and reductions in velocity that accompany reductions in flow result in reductions in heat transfer coefficients and chiller efficiency. There is still a net gain, and the exact amount varies, but a rule of thumb is that half of the pump energy savings potential is given up as chiller efficiency loss.

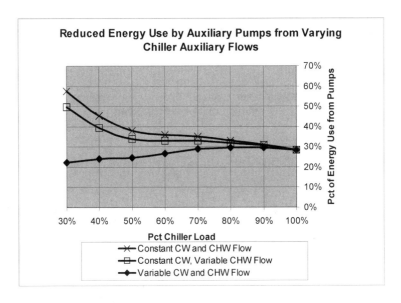

Figure 11-6. Chiller System Pump Energy Reduction from Variable Flow

COOLING TOWERS AND EVAPORATIVE FLUID COOLERS

This equipment uses the cooling effect of evaporation to cool a body of water.
- Over 90% of the capacity of a cooling tower or fluid cooler is from wet bulb temperature, regardless of dry bulb temperature. In dry climates, this effect can create dramatic temperature drops in the water, but in high humidity climates the "wet bulb depression" can be small.
- A cooling tower evaporates a portion of the water to cool the remaining water.
- A fluid cooler cools indirectly through a heat exchanger, so the cooling and cooled fluids do not touch. The heat exchanger in a fluid cooler creates a barrier for heat transfer and raises the final 'cooled' tempera-

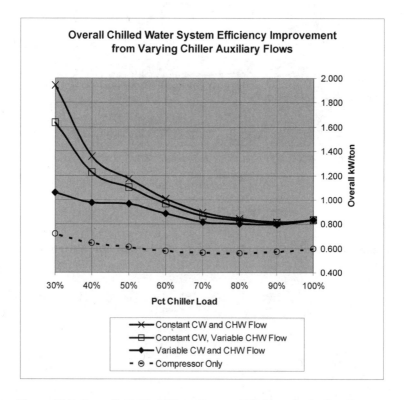

Figure 11-7. Overall Chilled Water System Efficiency Including Pumps
Assumes 0.6 kW/ton chiller at full load, and assumes half of the pump savings are given up to reduced chiller efficiency from reduced flow velocities.

Mechanical Systems 233

ture by 5-10 degrees, however are still widely used because the cooled fluid can be a sealed system without distributing the mineral and corrosion properties of standard cooling tower water; heat transfer surface cleaning is limited to the fluid cooler heat exchanger.

Cooling Tower Approach

A measure of capacity of both cooling towers and fluid coolers is "approach," which is how close the leaving water temperature is to the theoretical value of wet bulb temperature. For example, a cooling tower with a 10 degree approach at 60 degrees wet bulb will produce 70 degree leaving water.

- Lower approach temperatures are achieved either through increased water-air contact surface area or increased air flow.
- Increased air flow results in higher fan motor horsepower, so energy conservation would favor the use of larger body equipment to lower approach temperature instead of larger motors.
- Approach temperatures of 10 degrees are easy to achieve. Approach temperatures lower than 7 degrees result in exponential increases in cooling tower cost and are usually not warranted.
- The efficiency measure of a cooling tower is a combination of approach and fan motor horsepower. For example, a cooling tower used for a centrifugal chiller will be matched in terms of tons capacity. Actually, the cooling tower "tons" capacity includes an extra measure of heat for the heat of compression and may use 15,000 Btuh/ton instead of 12,000 Btuh/ton.
- A rule of thumb for cooling towers serving chillers is to be an order of magnitude less in specific power consumption.
- Approach temperatures higher than 10 degrees or fan motors higher than 0.05 kW/ton represent lost opportunities in leveraging water cooled technology benefits.
- Using 0.5 kW/ton as a benchmark for a chiller, a performance specification for an energy efficient cooling tower would be 0.05 kW/ton, at 7 degrees F approach to local design wet bulb.

DRY COOLERS

These are simple air coils. Normally filled with glycol for freeze protection, these units are simple and inexpensive. However, their performance is completely dependent on ambient dry bulb temperature.

A rule of thumb for leaving fluid temperature of a dry cooler is normally "Ambient Plus 30."

The energy implication of a dry cooler lies in the equipment it serves. For example, if a water-cooled refrigeration unit is connected to a dry cooler, it will see 110 degree condensing medium when it is 90 degrees outside. The energy penalty is significant compared to water that is cooled by evaporation in response to wet bulb temperature. The energy penalty for using 110 degree water instead of 70 degree water is 40-60%.

ELECTRONIC EXPANSION VALVES

Conventional expansion valves are normally sized using a 100 psi differential pressure to allow proper metering without hunting and over-feed that can damage compressors.

Electronic expansion valves can operate with less of a differential pressure across the valve without hunting and over-feeding. Eliminating this 'need' to have a high differential pressure allows the use of colder condenser water when it is available.

By replacing conventional expansion valves with electronic expansion valves, the lowest allowable operating head pressure requirement will be reduced. This measure can then be combined with reducing condenser water temperature to reduce condensing pressure and compressor power.

For example, the use of an electronic expansion valve may allow the use of 50 degree water for the condenser instead of 70 degree water, lowering compressor power 20-30 percent.

AIR AND WATER CIRCULATING SYSTEM RESISTANCE

Air Horsepower
The basic relationship for air horsepower is
$HP(air) = CFM * TSP/6356$
Where CFM is cubic feet per minute air flow and TSP is the total static pressure, in. w.c.

Water Horsepower
The basic relationship for water horsepower is
$HP(water) = GPM * HEAD/3960$
Where GPM, is gallons per minute water flow and HEAD is the total resisting pressure, ft. w.c.

NOTE: Motor Hp is higher than air/water horsepower. Brake horsepower (Bhp) includes the losses from fan/pump inefficiency. Motor horsepower (Mhp or Hp) represent the input power and energy, since they include the Bhp of the fan/pump, any drive losses (belt or coupling) and motor inefficiencies.

If the static pressure (TSP) or head loss (HEAD) reduction is known, savings can be estimated directly. For fittings and entrance/exit losses, dynamic factors are used and are referenced in tables published after empirical testing. These are in the form of "C" factors for air and "K" factors for water, and will vary depending upon the geometry of the fitting, the particulars of the entrance/exit condition, etc. In both cases, the dynamic losses are expressed in terms of "the number of velocity heads." The "velocity head" value multiplied by the "C" or "K" value is the anticipated loss of the item.

Velocity Head (Hv) Formulas
$Hv\ (air) = (V/4005)^2$
where Hv is in. w.c.
and V is velocity in feet per minute

$Hv\ (water) = (V^2/32.2)$
where Hv is ft. w.c.
and V is velocity in feet per second

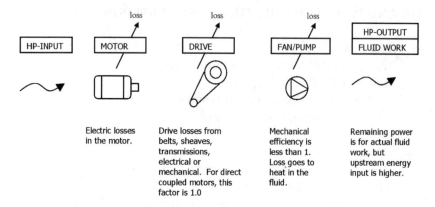

Figure 11-8. Fan/Pump Motor Work Diagram

The power required by the motor is greater than the fluid work because the pump or fan, drive connection and motor all incur losses. The fluid work must be divided by the product of the efficiencies of all losses along the way to determine the input requirement.

FAN/PUMP MOTOR WORK EQUATION

$$Hp\text{-}input = (Hp\ output)/(Eff_{fan} \times Eff_{drive} \times Eff_{motor})$$

Example: after determining the air horsepower is 0.5 hp, find the input power given fan efficiency is 75%, drive efficiency is 96%, and motor efficiency is 91%.

Ans: Input = 0.5/(0.75 * 0.96 * 0.91) = 0.76 hp

Understanding the multiple factors that affect input energy will help yield more accurate calculations. But it is also a reminder that energy reduction gains can come from process improvements (less fluid flow, less resistance), a more efficient fan or pump, drive improvements, or driver improvements (motor).

Mechanical Systems 237

FAN AND PUMP EFFICIENCIES, AND BELT DRIVE EFFICIENCIES

Device	Peak Eff. Range, %
Centrifugal Fan	
AF - Airfoil	75-85
BI - Backward Inclined	70-80
FC - Forward Curved	60-65
Axial Fan	
Vane Axial	75-85
Tube Axial	65-75
Propeller	45-50
Centrifugal Pump	
End Suction	65-75
Double Suction	75-85

Figure 11-9. HVAC Fan and Pump Typical Efficiencies

Motor Hp	Efficiency
1 and smaller	90%
2	92%
5	94%
10-50	95%
100 and larger	96%

11-10. Belt Drive Typical Efficiencies, Standard V-belts

FAN AND PUMP THROTTLING METHODS

Relative power requirements for throttling with discharge damper, inlet vanes, and VFD.

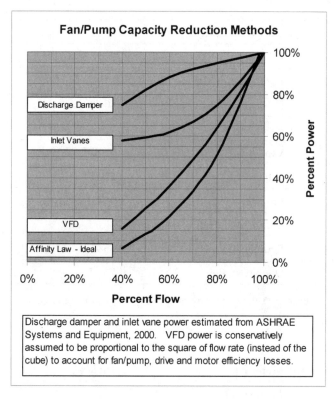

Figure 11-11. Power Requirements for Different Fan Throttling Methods

THERMAL BREAK-EVEN CONCEPT FOR BUILDINGS

Buildings are like boxes. They have thermal losses and gains through the shell (envelope), and they also have appliances and activities inside that generate heat. There will be a point where the envelope loss just matches the internal heat gains and neither heating nor cooling is required; above this break-even temperature cooling will be required, below this temperature heating will begin to be needed. This is generalized, and it is understood that there can be special areas requiring heating or cooling regardless of the bulk behavior of the building.

Implications of the Thermal Break-even Temperature:
Outdoor Air Economizer Savings
The lower the break-even temperature, the more hours there are where cooling is needed indoors while cool air is available 'for free' outdoors at the same time. These are economizer hours.

Supply Air Reset for VAV Savings
The higher the break-even temperature, the sooner VAV air flows will reach minimum and heating will become engaged for indoor comfort. Thus, for supply air reset based on outside air, the break-even point determines when it will be economical to raise supply air temperature. Raising supply air temperature saves energy by reducing reheat coil loads (simultaneous heating and cooling).

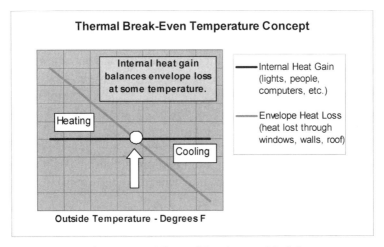

Figure 11-12. Thermal Break-even Model

AIR-SIDE ECONOMIZER

The economizer is most appropriate for thermally massive buildings which have high internal loads and require cooling in interior zones year round. **It is ineffective in thermally light buildings and buildings whose heating and cooling loads are dominated by thermal transmission through the envelope.** Economizer cycles will provide the greatest benefit in climates having more than 2000 (heating) degree days per year, since warmer climates will have few days cold enough to permit the use of outside air for cooling.

Source: *Energy Management Handbook* 6th Ed, Turner/Doty

The lower the break-even point the greater the benefit of the air economizer. The key to savings in an air economizer is the number of hours that outside air is cold enough to provide cooling that occur simultaneously with the need for cooling indoors. Buildings with cooling loads dominated by envelope offer little opportunity for economizer benefit, simply because when it is cold outside it is also cold inside. But buildings with high internal loads that dominate envelope loads, the economizer is viable. Below the "break even" temperature no cooling is needed. A further limitation of the air economizer is the humidity level of the outside air used for cooling. For most comfort applications, any humidity in the outside air is acceptable if the air temperature is 55 degrees or lower. If the dew point of the incoming air is too high, indoor rH will increase with opportunities for discomfort and building detriment. So, there is an economizer "cutoff point" above which the system will revert to mechanical cooling.

The higher the cutoff point the more economizer hours and savings. However, raising this should be done with caution to prevent high humidity issues indoors.

The outdoor temperatures bounded by the building break-even point and the economizer cut-off point are economizer hours. The HVAC load that occurs at these temperatures is usually low, but still represents avoided run time of the refrigeration equipment and makes the economizer worthwhile since it is very inexpensive to implement.

Mechanical Systems 241

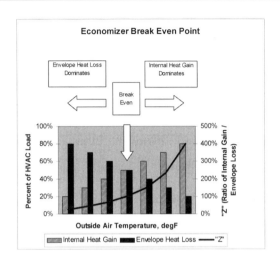

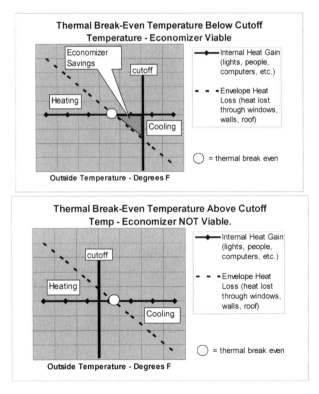

Figure 11-13. Effect of Thermal Break-even Temperature on Economizer Savings

Representative Values Only

Building Type	Internal Gains or Envelope Dominates?	Approximate Break Even Temp (estimates)	Air-Economizer Viable?
General - Residential	Envelope	65	No
General – Buildings with a low percentage of perimeter surface area (e.g. cube shaped)	Internal Gains	Varies	Yes
General – Large building core areas	Internal Gains	Varies	Yes
General – Buildings with a large amount of perimeter envelope area compared to interior area.	Envelope	65	No
Apartments, Condominiums, Hotel Guest Rooms	Envelope	65	No
Assembly area with high people density (church, theatre, ballroom)	Internal Gains	20-40	Yes
Church building, other than busy times	Envelope	65	No
Office building interior areas	Internal Gains	40-50	Yes
Office building perimeter areas, near the glass, standard construction	Varies	50-60	Marginal
Office building perimeter areas, near the glass, high performance glazing and well insulated.	Internal Gains	40-50	Yes
Restaurant – Kitchen	Internal Gains	30-40	Yes
Restaurant – Dining Area with windows	Envelope	65	No
Warehouse that is heated and cooled, but used just for storage	Envelope	65	No
School class rooms, standard room filled with students	Internal Gains	40-50	Yes
Light manufacturing, low activity, minor interior equipment loads, sparse people loading.	Envelope	55-65	Marginal
Hot Process Manufacturing, Ovens, Baking, Cooking, etc.	Internal Gains	10-30	Yes
Computer data center, 40W/SF	Internal Gains	0	No. Most Computer Equipment is Humidity-Sensitive

Figure 11-14. Break-even Temperatures for Different Building Activities

Degrees F	Dew point	Relative Humidity
OA Temperature Degrees F	OA OA	OA
70 deg	47 deg	44%
69 deg	47 deg	45%
68 deg	47 deg	47%
67 deg	47 deg	49%
66 deg	47 deg	50%
65 deg	47 deg	52%
64 deg	47 deg	54%
63 deg	47 deg	56%
62 deg	47 deg	58%
61 deg	47 deg	60%
60 deg	47 deg	62%
59 deg	47 deg	64%
58 deg	47 deg	66%
57 deg	47 deg	69%
56 deg	47 deg	72%
55 deg	any	any

Figure 11-15. Maximum Relative Humidity for use in Air Economizer
Based on maintaining indoor air conditions at 47 deg F dew point or less.

Thermal Break-Even Temperature, Degrees F	24x7 55 deg cutoff	6a-6p 55 deg cutoff
65 and higher	0	0
55-65	0	0
50-60	3-0%	2.5-0%
40-50	13-3%	10-2.5%
30-40	23-13%	19-10%
20-40	29-13%	25-10%
10-30	34-23%	29-19%

Figure 11-16. Air Economizer—Representative Savings
Cutoff 55 degrees. Calculated with Colorado Springs Weather Data. Savings are approximated.

COMPUTER ROOM AIR CONDITIONING (DATA CENTER)

Computer rooms/data centers are uniquely steady in load shape, weather independent, all sensible cooling, and humidity sensitive. Calculations for computer room energy savings calculations are often easy since the loads are steady. The largest electric loads are ones that practically nothing can be done about. The loads that can be influenced are HVAC and lighting. Lighting loads add directly to the cooling load so they should be as low as is practical. HVAC cooling energy is roughly 20-30% of the total electric load, depending on the cooling efficiency, so cooling efficiency is a natural target. This is usually achieved by lowering refrigeration condensing head pressure or raising refrigeration suction pressure or both. Humidification is very energy intensive, especially since there will always be simultaneous humidification and dehumidification and since these units are normally electric powered. Interestingly, the largest humidification demand comes from the unintentional dehumidification caused by the cooling coils—in many systems 10-15% of the cooling energy spent is on unintended dehumidification. Therefore the cooling apparatus dew point is a natural target, by either lowering the space rH, or by raising the chilled water temperature if chilled water is used.

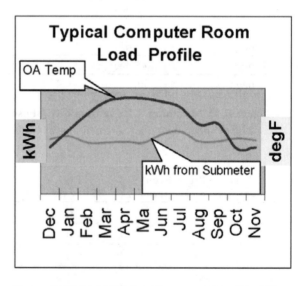

Figure 11-17. Data Center—Representative Load Shape vs. Outside Air Temperature
Source: "Energy Efficiency in Computer Data Centers," *Energy Engineering Journal*, Vol. 103, No. 5, 2006. Data Courtesy of Agilent Inc.

Data Center Load is nearly weather-independent

Mechanical Systems

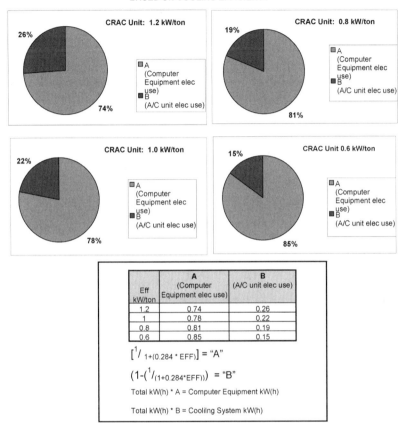

Figure 11-18. Proportions of Cooling Energy to Total Computer Room Energy
Source: "Energy Efficiency in Computer Data Centers," Doty, *Energy Engineering Journal*, Vol. 103, No. 5, 2006.

COOLING ENERGY BALANCE
FOR HEAT PRODUCING EQUIPMENT

To balance, Q=Q. Once the Q of the equipment is known (the heat loss), the A/C cooling Q must be equal to that. The electrical input to the cooling system is then a function of the COP of the cooling system. COP = EER/3.413. The process heat load in Btu then converts to cooling input Btu, and finally to cooling electric input.

Process heat X Factor = cooling input (same units)

Example: 100 kW of oven heat, cooled by EER-10 equipment, 100*0.34 = 34 kW cooling input

Example: 500,000 Btu of oven heat, cooled by EER-12 equipment, 500,000*0.28 = 140,000 Btu input = 41 kW.

EER	kW/ton	COP	FACTOR
8	1.5	2.344	0.4266
10	1.2	2.930	0.3413
12	1.0	3.516	0.2844
14	0.86	4.102	0.2438
16	0.75	4.688	0.2133

Figure 11-19. Factors for Cooling Energy Input When Paired to Known Heat Loads
Process heat * Factor = cooling input (same units)

HUMIDIFIERS

Distribution method (pan, fog, atomizing, and spray) is irrelevant to energy use. The way the moisture is converted to vapor is what is important. When electric resistance or infrared light emission is used to make steam, the energy use is directly calculated at 1000 Btu/lb.; for gas heating, the calculation is the same other than flue losses and the difference in fuel cost. Adiabatic (evaporation) humidification uses an order of magnitude less energy $(1/10^{th})$ for each unit of moisture compared to infrared or boiling.

Adiabatic humidifiers such as evaporative pads, atomizers, or ultrasonic get the heat of vaporization from the air and therefore cool and humidify the air simultaneously. Where simultaneous cooling and humidification are needed, such as in computer rooms; this can be leveraged for good savings, with the free cooling effect. However, if the cooling is not beneficial, the energy saved in the humidifier will be equal to the heat added to the air stream and the energy use will be equal.

Evaporative pad humidifiers are the least energy intensive of all, with only the air friction as a cost. Compressed air nozzle fogging humidifiers use considerable quantities of air (**See Figure 11-20**) and, with the compressor energy considered, are on par with ultrasonic. Ultrasonic humidifiers and high quality fog systems require de-ionized (DI) water and those costs must be considered.

Data Center Application
The following is taken from: "Energy Efficiency in Computer Data Centers," Doty, *Energy Engineering* Vol. 103, No. 5, 2006

Excess dryness encourages electrostatic discharge (ESD) and computer hardware problems, therefore, some humidification will normally be needed for data centers to reduce static and sparks. Depending on the equipment manufacturer, 30-35% rH may be enough to prevent spark issues; humidifying to 50% rH should be discouraged. Since humidification consumes energy, raising humidity levels higher than necessary should be avoided, and lowering humidity levels is an opportunity to reduce energy use.

[Author note: In one large data center, changing set points to achieve 30%rH instead of 45%rH allowed multiple humidifiers to turn off, reducing electric load by 70kW.]

The need for humidification, ironically, is due mostly to the natural dehumidification effect of the cooling coils. So the first thing to do is reduce the dehumidification. This is done by raising the cooling coil apparatus dew point by:
- Lowering the room dew point by increasing temperature and lowering relative humidity, after verifying that the environmental conditions meet the requirements of the equipment it serves.
- Increasing coil size so the average coil surface temperature is elevated. Although not currently a standard offering, it may be possible to request a DX computer room unit with a mismatched com-

pressor/evaporator coil pair, e.g. a 20-ton compressor and a 30-ton coil. Increasing coil size is done at the design stage, and may increase unit size and cost.
- Increasing chilled water temperature for chilled water systems. This is an integrated design choice also, and requires that the needed heat transfer be verified through coil selection to be available at the higher temperature entering water, e.g. 50 deg F entering Chilled Water (CHW). Increasing chilled water temperature is a design decision also, and will steer the design to a segregated system whereby the computer room chilled water is operated at a higher temperature than the building systems, through either a separate system entirely or a heat exchanger and blending valve arrangement. Verifying coil capacity meets heat loads with higher chilled water temperature is a prerequisite to this measure.
- Adjusting controls for wider tolerance of "cut-in" and "cut-out" settings, allowing indoor humidity levels to swing by 10% rH or more. The savings from this measure come from reduced control overlap between adjacent CRAC units, i.e. the controls of one machine calling for humidification while the controls of the neighboring machine calling for de-humidification.
- Coordinating the unit controls to act more like "one big CRAC unit" than multiple independent CRAC units. The savings from this measure are similar to widening the control settings, which are from reduced control overlap between adjacent CRAC units. The standard use of multiple CRAC units, each with their own "stand-alone" controls, each with tight tolerance control settings, is a built-

Medium	Technology	Energy Consumption	Remarks
Steam	Infrared Lamps	30+ kWh	
	Electric Resistance	30 kWh	
	Gas-Fired Steam	1.25 therms at 80%e	
Adiabatic	Compressed air atomization	3 kWh 12-15 cfm	Note 1
	Ultrasonic	2.5 kWh	Note 1
	Evaporative Pads	---	Note 1

Table Note 1: Adiabatic cooling can either help (if cooling is needed) or will require 1000 Btu of auxiliary heating per lb of evaporated moisture to compensate

Figure 11-20. Approximate Humidifier Energy Use per 100 lbs of Moisture

Mechanical Systems 249

in opportunity for simultaneous heating/cooling and humidification/de-humidification. The overlapping controls are readily observed in the field and function, but with energy penalty. If such overlap can be avoided, energy savings will result. Note: depending upon computer hardware heat density, there will naturally be different conditions at different CRAC units, so independent temperature control at each CRAC unit is appropriate.
- Increasing air flow to raise average coil surface temperature and air temperature. Note that this measure increases fan energy sharply and may exceed the avoided humidification savings.

KITCHEN GREASE HOODS (TYPE 2)

Kitchen Hood Exhaust

Field constructed hoods are governed by building codes for air flow, such as 50 CFM/SF for wall-hoods and 75 CFM per SF for island hoods.

U.L. Listed "engineered" hoods use approximately 30-40% less air compared to field constructed hoods. So, for this example, a wall hood would use 30 CFM per SF and an Island hood would use 45 CFM per SF if they are an engineered hood. These hoods are an engineered package with specified air flows for exhaust and make-up stamped on the hood.

Kitchen Hood Make-up

The source of savings for kitchen hoods is to reduce the exhaust quantity if possible, since that also reduces make up air, and to temper the make up air as little as possible to achieve reasonable comfort, recognizing that any tempering of the air is thrown away.
- Turning off the hoods when not cooking saves energy. Often the hoods are started at 6am out of habit, or are used to cool the kitchen.
- Make up as much as practical right at the hood so conditioned air from the space is spared. Make up for the hood can be in front of the hood, at the lip of the hood, or short circuiting within the hood. Make up percent can be 0 up to 80%, with 60% being common.

HEAT PUMPS

These utilize the vapor compression cycle and equipment, but the "useful" portion of the cycle is generally the heat, which makes it reverse of standard refrigeration where the heat is rejected.

Like all refrigeration cycles, the work is determined by the lift of the cycle (difference between low and high pressures). For systems using air as a source of heat, this is an immediate disadvantage since you need the most heating when it is coldest outside and that is when the efficiency is least. Air systems also develop a coating of ice on the outdoor coils and must be periodically defrosted by a temporary reversal of the cycle and through auxiliary heating.

Heat pumps are normally used for cooling in summer, via a refrigerant "reversing valve." Using the same apparatus for both heating and cooling presents a formidable design challenge since the heating and cooling are almost invariably different. The result is sizing for the dominant load and having excess capacity in the opposite season. Short cycling can result and some form of capacity reduction (2-speed compressors, etc.) will usually be needed.

In cases where one half of the cycle is rejected (cooling or heating) efficiencies are the same as refrigeration equipment, plus or minus the effect of the ambient temperatures. However, in cases where the heating and cooling are BOTH used and neither is rejected the efficiencies are much higher.

Air-to-air Heat Pumps (Air-source Heat Pumps)

These are more efficient than gas heating in mild weather (40-65 degrees), but the efficiency drops rapidly with temperature and below 25 degrees are essentially electric resistance heaters below 25 degrees F. Viability is determined by annual hours of use and corresponding outdoor temperatures. Connected load is higher due to coincident compressor and electric heat operation in defrost mode.

An ideal, but probably impractical, system in moderate climates would utilize the heat pump down to 40 degrees and only use gas heating below that.

Using weather data of average temperatures and corresponding hours, savings benefit can be calculated. Assuming a 40 degree break-even point (where electric heating takes over) and a building where heating is required below 65 degrees, the following example chart shows the times of the year that an air source heat pump will have an efficiency benefit over straight electric heat. This example is for a hotel with standard PTACs that

Mechanical Systems

	Jan	Feb	Mar	Apr	May	Jun	Jul	Aug	Sep	Oct	Nov	Dec
0000	25.2	29.2	40.6	50.6	59.6	66.6	69.6	69.6	69.6	53.6	43.2	31.2
0100	23.9	27.9	39.3	49.3	58.3	65.3	68.3	68.3	62.3	52.3	41.9	29.9
0200	22.7	26.7	38.1	48.1	57.1	64.1	67.1	67.1	61.1	51.1	40.7	28.7
0300	21.7	25.7	37.1	47.1	56.1	63.1	66.1	66.1	60.1	50.1	39.7	27.7
0400	20.9	24.9	36.3	46.3	55.3	62.3	65.3	65.3	59.3	49.3	38.9	26.9
0500	20.7	24.7	36.1	46.1	55.1	62.1	65.1	65.1	59.1	49.1	38.7	26.7
0600	21.2	25.2	36.5	46.6	55.6	62.6	65.6	65.6	59.6	49.6	39.2	27.2
0700	22.4	26.4	37.8	47.8	55.8	63.8	66.8	66.8	60.8	50.8	40.4	28.4
0800	24.7	28.7	40.1	50.1	59.1	66.1	69.1	69.1	63.1	53.1	42.7	30.7
0900	27.9	31.9	43.3	53.3	62.3	69.3	72.3	72.3	66.3	56.3	45.9	33.9
1000	31.7	35.7	47.1	57.1	66.1	73.1	76.1	76.1	70.1	60.1	49.7	37.7
1100	35.9	39.9	51.3	61.3	70.3	77.3	80.3	80.3	74.3	64.3	53.9	41.9
1200	39.9	43.9	55.3	65.3	74.3	81.3	84.3	84.3	78.3	68.3	57.9	45.9
1300	42.9	46.9	58.3	68.3	77.3	84.3	87.3	87.3	81.3	71.3	60.9	48.9
1400	44.9	48.9	60.3	70.3	79.3	86.3	89.3	89.3	83.3	73.3	62.9	50.9
1500	45.6	49.6	61.0	71.0	80.0	87.0	90.0	90.0	84.0	74.0	63.6	51.6
1600	44.9	48.9	60.3	70.3	79.3	86.3	89.3	89.3	83.3	73.3	62.9	50.9
1700	43.1	47.1	58.5	68.5	77.5	84.5	87.5	87.5	81.5	71.5	61.1	49.1
1800	40.4	44.4	55.8	65.8	74.8	81.8	84.5	84.8	78.8	68.8	58.4	46.4
1900	37.1	41.1	52.5	62.5	71.5	78.5	81.5	81.5	75.5	65.5	55.1	43.1
2000	33.9	37.9	49.3	59.3	68.3	75.3	78.3	78.3	72.3	62.3	51.9	39.9
2100	31.2	35.2	46.6	56.6	65.6	72.5	75.8	75.6	69.6	59.6	49.2	37.2
2200	28.7	32.7	44.1	54.1	63.1	70.1	73.1	73.1	67.1	57.1	46.7	34.7
2300	26.7	30.7	42.1	52.1	61.1	68.1	71.1	71.1	65.1	55.1	44.7	32.7

Air Source Heat Pump Hours of Operation (Denver Area)
assumes heating is required below 65 degrees F outside air
assumes electric resistance heating takes over at 40 degrees

Hours of operation between 40 and 65 degrees, when heat pump would be used.
Hours of operation below 40 degrees when electric resistance heaters would be used.

Figure 11-21. Air-Source Heat Pump benefit at different temperatures.

can consider the heat pump option upon replacement. Detailed analysis would be done in spreadsheet form using bin hours and consider efficiency changes at various temperatures. A similar analysis can be performed with heat pump vs. gas heat for standard furnaces.

Water-to-air Heat Pumps (Water-source Heat Pumps)
Depending on the source of the water, these can be very useful. If, for example, a continuous source of tepid water exists with no energy input, the water supply would provide an excellent source and sink.

A common HVAC system utilizing water source heat pumps connects a number of heat pumps to a common circulating system. During moderate weather when some areas need cooling and some areas need heating this system is extremely efficient, because it is just moving the heat around the building. But during hot summer and cold winter months, this system is very energy intensive because when all the heat pumps are in one mode or the other, additional energy must be expended to stabilize the loop temperature.

Water-to-water Heat Pumps
These have varied application. Applied to dedicated cooling areas using "house" condenser water, these have efficiency consistent with water-cooled equipment.

Excellent COPs where there exists a coincident and adjacent need for cooling and heating. In these cases, NONE of the heat is rejected on either side and the overall COP is very high as a result. An example is a need for heating hot water and cooling at the same time; using the heat pump to provide all the heating and a portion of the cooling can provide 140 degree hot water at COP=2 with free cooling and an overall COP=4.

Ground-source Heat Pumps (GSHP)
These are a variation of a Water-to-Air heat pump, using buried piping as a heat source/heat sink. Very high efficiencies are achievable with this system due to the large heat source/sink the earth provides. COP of 4.0 and EER of 17 are common with this system. Care in design and life cycle costing is recommended, to avoid over-stating long term savings, since the ground has a finite heat capacity and will "creep" up or down in temperature depending on whether the summer or winter heat flux dominates. For most commercial facilities, the cooling load will dominate and thus the ground will be expected to heat up over time, unless some form of intervention is applied to prevent this (adding cooling towers) or mitigate this (over-sizing the loop to delay the effect).

REFRIGERATION CYCLE

The vapor compression cycle enjoys widespread use in HVAC cooling, HVAC heating, refrigeration, process cooling, as well as many ancillary uses. From an energy standpoint, the goal is always the same. Lower the head pressure or raise the suction pressure, where possible, to reduce compression "lift" and motor kW. The typical effect of either measure is 1-1.5% improvement per degree F raised.

The challenge is to find economical ways to lower head pressures and raise suction pressures, while not creating any adverse effects on the customer's business process or equipment.

The work of the cycle (kWh) is a function of the difference between the upper and lower horizontal lines in the Mollier Diagram.

Source of basic diagrams: Wikipedia

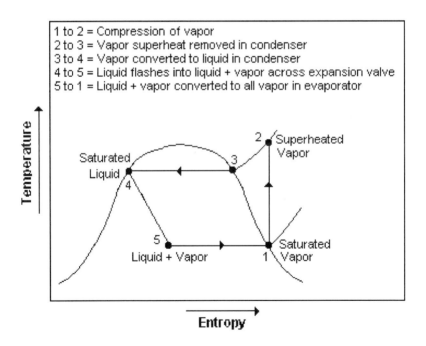

Figure 11-22. Basic Refrigeration Cycle

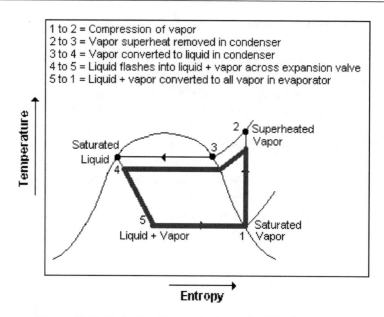

Figure 11-23. Reducing Energy by Lowering Head Pressure

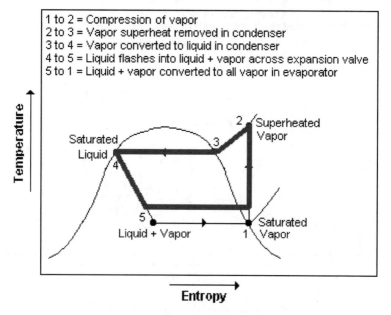

Figure 11-24. Reducing Energy by Raising Suction Pressure

Mechanical Systems

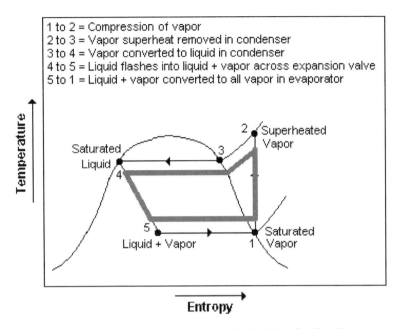

Figure 11-25. Lowering Head Pressure AND Raising Suction Pressure

EVAPORATIVE COOLING

Direct Evaporative Cooling

The air passes through the evaporative media and cools adiabatically, without a change in enthalpy. As the air absorbs moisture it cools proportionally so the total heat of the air remains unchanged. Since each subsequent pass adds moisture, this process is best suited for "single pass" applications, where the air is cooled/humidified and exhausted without recirculation. Attempts to recirculate air through a direct evaporation system risk high moisture issues in the space.

Applications for direct evaporation are kitchen hoods, factories, and other facilities where the air can be exhausted after a single pass.

Direct evaporation can also be used to advantage in air heat recovery equipment to lower the exhaust temperature and amplify the heat recovery effect without adding moisture.

The temperature of the air leaving the evaporative cooling equipment is a function of the ambient wet bulb temperature and thus the use of this technology is limited to climates where the ambient wet bulb temperature is sufficiently below the desired dry bulb supply air temperature.

Supply air for direct evaporative cooling systems is usually at somewhat higher temperature than achieved with mechanical refrigeration cooling, such as 60-62 degrees F. Consequently, increased air flow rates (cfm) are needed with correspondingly larger duct systems and fans.

The relationship for predicting direct evaporative air cooler performance is:

LATdb degrees F= EATdb- (eff*(EATdb—AMBwb))

Where:
LATdb = leaving air temperature, deg F dry bulb
EATdb = entering air temperature, deg F dry bulb
Eff = efficiency of the evaporative cooler. Range is 0.5 for felt pads to 0.9 for 12 inch thick media)
AMBwb = ambient wet bulb temperature, deg F.

Example, entering air is 65 degrees, ambient wet bulb temperature is 50 degrees, and the equipment uses 6 inch media at 80% efficiency.

Leaving air temperature = 65 – (0.8 * (65–50)) = 53 deg F leaving air temperature.

Indirect-direct Evaporative Cooling

Two sequential evaporative cooling processes provide improved cooling with less added moisture.

The first stage uses indirect cooling via a heat exchanger, sensibly cooling the air. Schematically the first stage is a cooling tower that produces water within 7-10 degrees of the ambient wet bulb temperature. The cooled water is circulated through a coil and cools it to a mid range temperature.

The second stage is a conventional direct evaporative cooling unit.

The advantage to adding the indirect step is to begin the adiabatic cooling at a lower initial temperature, thereby adding less moisture. By pairing this system with an appropriate fraction of constant building exhaust, the remainder of the air can be re-circulated without moisture damage indoors.

The use of this system is limited by ambient dew point temperatures—if operated during high dew point conditions the indoor relative humidity will rise and create discomfort for occupants.

Indirect-Direct Evaporative Cooling with Supplemental Conventional Cooling

A variation of this system uses supplemental cooling to extend the effect of the first stage (indirect) cooling, thereby allowing operation in the indoor comfort zone throughout a greater range of outdoor wet bulb conditions.

Direct Evaporative Cooling—Post Cooling for Cooling Coils

Basis of savings: Evaporative cooling augments mechanical cooling.

- Effective in many climate regions, when outdoor wet bulb temperatures are sufficiently low (below 60 deg F wet bulb).

- Many HVAC cooling coils deliver 55 deg F air at or near saturation, so if the mechanical cooling is allowed to bring the air temperature to a certain point and stop, the evaporative step can do the remaining portion of cooling work with less energy and end up at the same psychrometric point.

- Can reduce mechanical cooling burden by 25%

Psychrometric Diagrams of Evaporative Cooling Processes

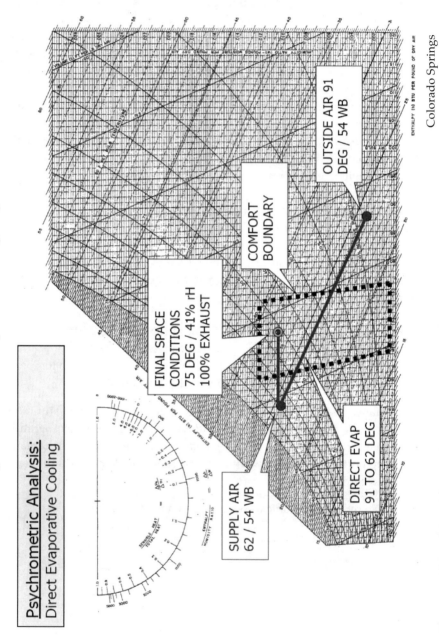

Figure 11-26. Direct Evaporative Cooling—Psychrometrics

Mechanical Systems 259

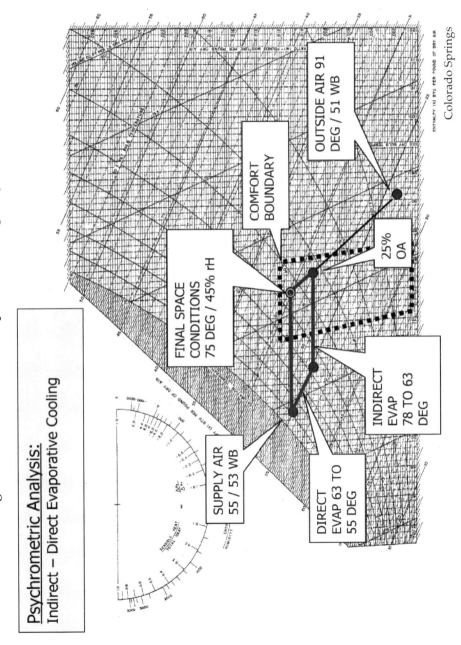

Figure 11-27. Indirect-direct Evaporative Cooling—Psychrometrics

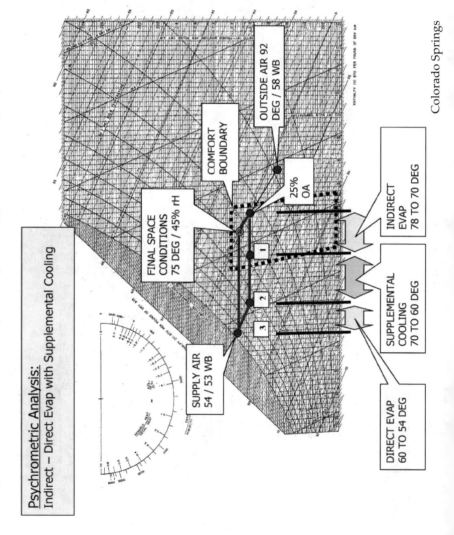

Figure 11-28. Indirect-direct Evap. Cooling with Supplemental Cooling–Psychrometrics

Mechanical Systems 261

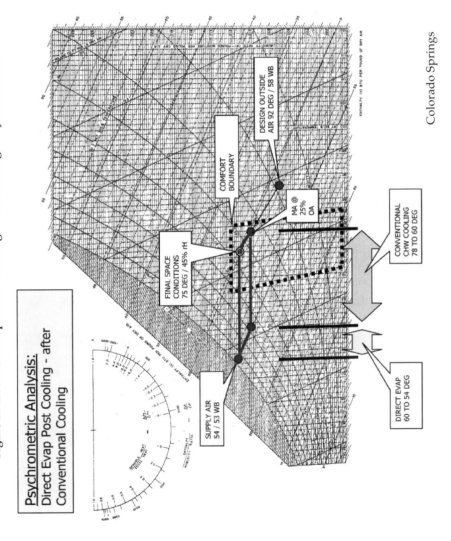

Figure 11-29. Direct Evaporative Cooling—Post Cooling—Psychrometrics

SPOT COOLING

If the dry bulb temperature of the air is below skin temperature, convection rather than evaporation cools workers. In these conditions, an 80 deg F air stream can provide comfort regardless of its relative humidity.

Source: ASHRAE Applications Handbook, 2003, © American Society of Heating, Refrigerating and Air-Conditioning Engineers, Inc., www.ashrae.org.

Comfort can be had at higher temperatures by using higher air flows, depending upon the relative humidity of the space. If the building temperature is raised 5 degrees from the benefit of strong air movement, the envelope losses will be reduced by 10% (2% per degree). NOTE: Cooling only the worker locations in a building, and not the building itself, will yield more savings.

Velocity, fpm	rH, %	Comfortable Temperature, deg F
0	20	76
0	40	75
0	50	74
0	60	73
300	20	81
300	40	80
300	50	79
300	60	78
600	20	85
600	40	83
600	50	81
600	60	80

Note: Consider air velocities in applications where papers could blow.

Figure 11-30. Air Velocity Effect on Comfort Zone

Data extracted from chart "Change in Human Comfort Zone as Air Movement Increases," ASHRAE Applications Handbook, 2003, © American Society of Heating, Refrigerating and Air-Conditioning Engineers, Inc., www.ashrae.org.

(Values are center of the range shown.)

Mechanical Systems 263

VAV REHEAT PENALTY

Whenever supply air is delivered below room temperature, and heating is required in the room, the heating load for that room is not the total heat required. The supply air must be heated as well. This is the VAV reheat penalty that is built into all single path VAV systems.

This can be mitigated through
- Supply air reset in heating season, although if the VAV system serves zones needing cooling this can create comfort issues.
- Reduced air flow minimum settings, provided sufficient ventilation is maintained
- Separating the ventilation air into a separate system of ductwork, making the main VAV system a recirculating system and allowing zero VAV box settings. The separate system is called dedicated outside air system (DOAS).

GLYCOL VS. EFFICIENCY

Glycol Effect on Energy Use

Basis of savings: Lower heat exchanger differential temperature to achieve the same heat exchange, lower system friction resistance through pipes, and higher pump efficiency.

- Glycol increases viscosity which retards heat transfer. In turn, higher heat exchanger approaches are required
  — For heating applications, the 'hot side' (e.g. flue gas) must be hotter to cause equivalent heat exchange.
  — For cooling applications, the 'cold side' (e.g. refrigerant) must be colder to cause equivalent heat exchange.
- Glycol effect is very pronounced in chilled water applications, and less pronounced with higher temperatures found in heating systems.
- Glycol viscosity increase adds pumping energy in several ways.
  — Adds system resistance for fluid in pipes, heat exchanger tubes, etc.
  — Reduces centrifugal pump efficiency.
  — If reduced heat exchange requires additional water flow, losses are compounded.

	Water to 30% EG	
Heat exchanger capacity	6.2 % decrease	Source: Sample tube and shell heat exchanger selection.

Figure 11-31. Glycol Effect on Heating Water Efficiency

	30% EG to water	30% PG to water	30% PG to 30% EG	Equipment Type
Chiller Power kW	3.6% decrease	6.8% decrease	3.4% decrease	Centrifugal Chiller - average
Chiller Power kW	1.6% decrease	2.5% decrease	0.9% decrease	Air Cooled Screw Chiller

Figure 11-32. Glycol Effect on Chiller Efficiency

All figures are at 45 degrees F chilled water.

Sources: Sample equipment selections from chiller equipment manufacturers

Note the differences between types of chillers; these have to do with whether the glycol is in the tubes or in the shell.

Note for viscosity charts:
EG = ethylene glycol
PG = propylene glycol

40 degF
Pump Selection Information
500 gpm, 24 ft with water (baseline)
head loss calculated with different viscosities

	ft head	bhp	SSU	ft2/sec
30% PG	31	5.8	44	72
30% EG	27	4.7	38	35
Plain Water	24	4.1	32	16

	savings
PG->EG	19.0%
PG->Water	29.3%
EG->Water	12.8%

0 degF
Pump Selection Information
500 gpm, 26 ft with water (baseline)
head loss calculated with different viscosities

	ft head	bhp	SSU	ft2/sec
45% PG	47	10.4	206	480
45% EG	34	7.5	66	128
Plain Water	26	4.3	36	26

	savings
PG->EG	27.9%
PG->Water	58.7%
EG->Water	42.7%

Summary of Pumping Penalty with Glycol (Bhp)

	40 degF	10 degF
PG->EG	19.0%	27.9%
PG->Water	29.3%	58.7%
EG->Water	12.8%	42.7%

Figure 11-33. Glycol Effect on Chilled Water Pumping Energy

Sources: Sample equipment selections from pump equipment manufacturers
Increase includes the viscous effect on pump efficiency as well as pipe friction loss. System calculation was for 500 gpm, 300 feet of pipe, 30 elbows using viscosity of water, EG, and PG at the same temperature. See **Figure 11-33**.

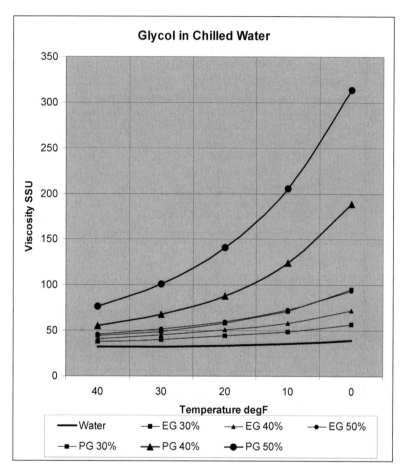

Figure 11-34. Viscosity Values for Glycol in Chilled Water
Units are SSU
Note: EG 40% and PG 30% lines are almost equal and are graphed on top of each other. Source: ITT Corp
Data for graph follows:

		40 degF	30 degF	20 degF	10 degF	0 degF
	Water	32	32	34	36	39
30%	EG	38	40	44	48	56
30%	PG	44	49	58	72	95
40%	EG	41	45	51	58	72
40%	PG	55	67	88	124	189
50%	EG	46	52	60	73	94
50%	PG	77	101	141	206	314

Mechanical Systems 267

COST OF VENTILATION

- Ventilation is necessary for a variety of reasons, and should not be reduced below what is necessary. However, excessive ventilation is costly and is a source of savings. Excess ventilation can be from:
  — Ventilation for occupants that aren't there.
  — Make-up air for exhaust fans that are left on for no reason
  — Unwanted exfiltration or infiltration through envelope leakage

- The energy consequence of the extra ventilation lies in the tempering of the air. In humid climates, the cost of ventilation also includes dehumidification which can be considerable.

Energy Consumption for Heating Outside Air
Units = therms
Table is based on 100 cfm of outside air at an average temperature being heated for 1000 hours to final temperatures shown.
Formula used: 1.1 * Altitude Factor * cfm * dT.
For gas heating, divide output by efficiency to obtain input.
For table use, determine average outside temperature during the period of interest.
Formula can be used directly with bin weather data if desired.

		Bldg Space Temp, degF		
		65	**70**	**75**
OA	**30**	39	44	50
Temp	**35**	33	39	44
degF	**40**	28	33	39
	45	22	28	33
	50	17	22	28
	55	11	17	22
	60	6	11	17

Figure 11-35. Energy Consumption for Heating Outside Air

SIMULTANEOUS HEATING AND COOLING

Overlapping heating and cooling occurs in many buildings. It can be deliberate or accidental. In all cases there is an energy penalty. A very constructive general action item is to identify and eliminate this waste. It is analogous to driving with the brakes on.

Larger HVAC systems that serve multiple zones tend to have more of this built into the buildings than smaller commercial HVAC, since smaller buildings are often served with unitary package HVAC equipment that is either heating OR cooling, but never both.

Some sources of this overlap are:
- Excessive minimum settings on VAV boxes
- Dehumidification cycle through over-cooling and reheating
- Unintentional overlap of cooling and heating set points so the heating and cooling are fighting
- Control sequenced heating and cooling coils that are in series
- Poorly adjusted electronic actuator open and close travel limits
- Valves that do not fully close off, either from system pressure, debris, or damaged metal seats
- Bare piping, especially for bypass legs of 3-way valves within an air stream
- Small area needing cooling at all times
- Certain occupants that who are always either too hot or too cold
- Space heaters
- Spot coolers
- Defective electric heat relays (stuck "on")
- Baseboard heat that is controlled independently of main HVAC systems
- Boilers left to idle in summer and circulate hot water through the building
- Chillers left to idle in winter and circulate chilled water through the building
- Comfort zoning problems that are fixed with duct heating coils.

Chapter 12

Motors and Electrical Information

GENERAL

Motor-driven equipment accounts for 64% of the electricity used in the U.S. industrial sector.
Source: US DOE Office of Energy Efficiency and Renewable Energy.

hp	Open Drip-Proof					
	1200 RPM (6-pole)		1800 RPM (4-pole)		3600 RPM (2-pole)	
	EPAct-1992	NEMA Premium	EPAct-1992	NEMA Premium	EPAct-1992	NEMA Premium
1	80	82.5	82.5	85.5	N/A	77
1.5	84	86.5	84	86.5	82.5	84
2	85.5	87.5	84	86.5	84	85.5
3	86.5	88.5	86.5	89.5	84	85.5
5	87.5	89.5	87.5	89.5	85.5	86.5
7.5	88.5	90.2	88.5	91	87.5	88.5
10	90.2	91.7	89.5	91.7	88.5	89.5
15	90.2	91.7	91	93	89.5	90.2
20	91	92.4	91	93	90.2	91
25	91.7	93	91.7	93.6	91	91.7
30	92.4	93.6	92.4	94.1	91	91.7
40	93	94.1	93	94.1	91.7	92.4
50	93	94.1	93	94.5	92.4	93
60	93.6	94.5	93.6	95	93	93.6
75	93.6	94.5	94.1	95	93	93.6
100	94.1	95	94.1	95.4	93	93.6
125	94.1	95	94.5	95.4	93.6	94.1
150	94.5	95.4	95	95.8	93.6	94.1
200	94.5	95.4	95	95.8	94.5	95

Figure 12-1. Motor Efficiencies
Nominal motor efficiency for NEMA Induction Motors Rated 600 Volts or Less
Source: Copper Development Association Inc., www.copper.org

hp	Totally Enclosed Fan-Cooled					
	1200 RPM (6-pole)		1800 RPM (4-pole)		3600 RPM (2-pole)	
	EPAct-1992	NEMA Premium	EPAct-1992	NEMA Premium	EPAct-1992	NEMA Premium
1	80	82.5	82.5	85.5	75.5	77
1.5	85.5	87.5	84	86.5	82.5	84
2	86.5	88.5	84	86.5	84	85.5
3	87.5	89.5	87.5	89.5	85.5	86.5
5	87.5	89.5	87.5	89.5	87.5	88.5
7.5	89.5	91	89.5	91.7	88.5	89.5
10	89.5	91	89.5	91.7	89.5	90.2
15	90.2	91.7	91	92.4	90.2	91
20	90.2	91.7	91	93	90.2	91
25	91.7	93	92.4	93.6	91	91.7
30	91.7	93	92.4	93.6	91	91.7
40	93	94.1	93	94.1	91.7	92.4
50	93	94.1	93	94.5	92.4	93
60	93.6	94.5	93.6	95	93	93.6
75	93.6	94.5	94.1	95.4	93	93.6
100	94.1	95	94.5	95.4	93.6	94.1
125	94.1	95	94.5	95.4	94.5	95
150	95	95.8	95	95.8	94.5	95
200	95	95.8	95	96.2	95	95.4

Figure 12-1. Motor Efficiencies (*Continued*)

VOLTAGE IMBALANCE

Voltage imbalance is undesirable for three phase motors since it causes a partial reverse rotation force which then acts like a brake. The extent of the imbalance determines the extent of the dragging and the efficiency loss. Rule of thumb is that imbalance should not exceed 1%.

It takes very little imbalance to de-rate a motor hp from nameplate, but it takes a lot of imbalance to do much to the efficiency.

$$\text{Voltage \% imbalance} = \frac{\text{(max voltage difference between any phase and the average voltage)}}{\text{Average voltage}}$$

Motors and Electrical Information

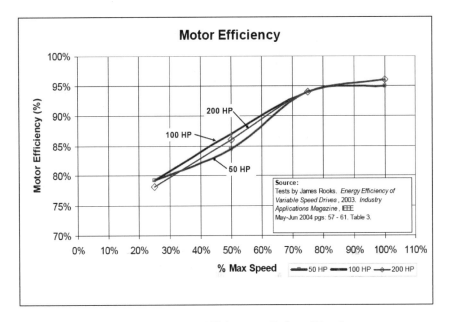

Figure 12-2. Motor Efficiency at Reduced Load
Source: Stum and Koran, "Techniques and Tips for Retro Commissioning Energy Calculations," Presentation at the National Conference on Building Commissioning, 2007. Original Data Source as noted.

Example:
470V	Voltage A-B (phase A to phase B)
460V	Voltage B-C
475V	Voltage A-C
468V	**Average voltage**
2V	A-B differential (voltage − average)
8V	**B-C differential**
7V	A-C differential
1.7%	Voltage Imbalance, from 8/468

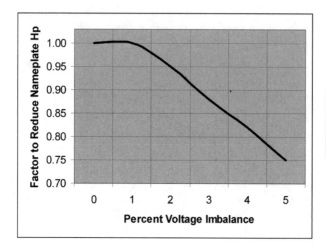

Figure 12-3. Voltage Imbalance Effect on Motor Hp Rating
Data Source: "Electric Motor Voltage Quality Problems," Industrial Technologies Program (ITP), 2006, US DOE Office of Energy Efficiency and Renewable Energy.
Original Source NEMA Standards Publication MG-1-2003, Motors and Generators.

Due to motor heating, motors with unbalanced voltage are to be reduced from nameplate maximum horsepower, according to this chart.

% Motor Load	Balanced	1% Imbalance	2.5% Imbalance
100	0.0%	0.0%	1.5%
75	0.0%	0.1%	1.4%
50	0.0%	0.6%	2.1%

Figure 12-4. Voltage Imbalance Effect on Motor Efficiency Rating
Source: "Energy Tips – Motor Systems," Industrial Technologies Program (ITP), 2005, US DOE Office of Energy Efficiency and Renewable Energy.
Note: the test data for this table was for an 1800 rpm 100 Hp motor, however the general trend of efficiency reduction with increased voltage unbalance is observed for all motors at all load conditions.

Losses	Fixed or variable loss	2-Pole average	4-Pole average	Factors affecting losses
Core losses	Fixed	19%	21%	Electrical steel, air gap, saturation
Friction and windage losses	Fixed	25%	10%	Fan efficiency, lubrication, bearings
Stator I^2R losses	Varies with load	26%	34%	Conductor area, mean length of turn, heat dissipation
Rotor I^2R losses	Varies with load	19%	21%	Bar and end ring area and material
Stray load losses	Varies with load	11%	14%	Manufacturing processes, slot design, air gap

Figure 12-5. Sources of Electric Motor Losses
Source: "Effect of Repair/Rewinding On Motor Efficiency," 2003, Electrical Apparatus Service Association (EASA)

Commercial Energy Auditing Reference Handbook

	KW Multipliers To Determine Capacitor KVAR Required														
	Corrected Power Factor														
		0.80	0.81	0.82	0.83	0.84	0.85	0.86	0.87	0.88	0.89	0.90	0.91	0.92	0.93	0.94
Original Power Factor	0.60	0.58	0.61	0.64	0.66	0.69	0.71	0.74	0.77	0.79	0.82	0.85	0.88	0.91	0.94	0.97
	0.61	0.55	0.58	0.60	0.63	0.65	0.68	0.71	0.73	0.76	0.79	0.81	0.84	0.87	0.90	0.94
	0.62	0.52	0.54	0.57	0.59	0.62	0.65	0.67	0.70	0.73	0.75	0.78	0.81	0.84	0.87	0.90
	0.63	0.48	0.51	0.53	0.56	0.59	0.61	0.64	0.67	0.69	0.72	0.75	0.78	0.81	0.84	0.87
	0.64	0.45	0.48	0.50	0.53	0.55	0.58	0.61	0.63	0.66	0.69	0.72	0.74	0.77	0.81	0.84
	0.65	0.42	0.45	0.47	0.50	0.52	0.55	0.58	0.60	0.63	0.66	0.68	0.71	0.74	0.77	0.81
	0.66	0.39	0.41	0.44	0.47	0.49	0.52	0.54	0.57	0.60	0.63	0.65	0.68	0.71	0.74	0.78
	0.67	0.36	0.38	0.41	0.44	0.46	0.49	0.51	0.54	0.57	0.60	0.62	0.65	0.68	0.71	0.75
	0.68	0.33	0.35	0.38	0.41	0.43	0.46	0.48	0.51	0.54	0.57	0.59	0.62	0.65	0.68	0.72
	0.69	0.30	0.33	0.35	0.38	0.40	0.43	0.46	0.48	0.51	0.54	0.56	0.59	0.62	0.65	0.69
	0.70	0.27	0.30	0.32	0.35	0.37	0.40	0.43	0.45	0.48	0.51	0.54	0.56	0.59	0.62	0.66
	0.71	0.24	0.27	0.29	0.32	0.35	0.37	0.40	0.43	0.45	0.48	0.51	0.54	0.57	0.60	0.63
	0.72	0.21	0.24	0.27	0.29	0.32	0.34	0.37	0.40	0.42	0.45	0.48	0.51	0.54	0.57	0.60
	0.73	0.19	0.21	0.24	0.26	0.29	0.32	0.34	0.37	0.40	0.42	0.45	0.48	0.51	0.54	0.57
	0.74	0.16	0.18	0.21	0.24	0.26	0.29	0.32	0.34	0.37	0.40	0.42	0.45	0.48	0.51	0.55
	0.75	0.13	0.16	0.18	0.21	0.24	0.26	0.29	0.32	0.34	0.37	0.40	0.43	0.46	0.49	0.52
	0.76	0.11	0.13	0.16	0.18	0.21	0.24	0.26	0.29	0.32	0.34	0.37	0.40	0.43	0.46	0.49
	0.77	0.08	0.10	0.13	0.16	0.18	0.21	0.24	0.26	0.29	0.32	0.34	0.37	0.40	0.43	0.47
	0.78	0.05	0.08	0.10	0.13	0.16	0.18	0.21	0.24	0.26	0.29	0.32	0.35	0.38	0.41	0.44
	0.79	0.03	0.05	0.08	0.10	0.13	0.16	0.18	0.21	0.24	0.26	0.29	0.32	0.35	0.38	0.41
	0.80		0.03	0.05	0.08	0.10	0.13	0.16	0.18	0.21	0.24	0.27	0.29	0.32	0.35	0.39
	0.81			0.03	0.05	0.08	0.10	0.13	0.16	0.18	0.21	0.24	0.27	0.30	0.33	0.36
	0.82				0.03	0.05	0.08	0.10	0.13	0.16	0.19	0.21	0.24	0.27	0.30	0.34
	0.83					0.03	0.05	0.08	0.11	0.13	0.16	0.19	0.22	0.25	0.28	0.31
	0.84						0.03	0.05	0.08	0.11	0.13	0.16	0.19	0.22	0.25	0.28
	0.85							0.03	0.05	0.08	0.11	0.14	0.16	0.19	0.22	0.26
	0.86								0.03	0.05	0.08	0.11	0.14	0.17	0.20	0.23
	0.87									0.03	0.05	0.08	0.11	0.14	0.17	0.20
	0.88										0.03	0.06	0.08	0.11	0.14	0.18
	0.89											0.03	0.06	0.09	0.12	0.15
	0.90												0.03	0.06	0.09	0.12
	0.91													0.03	0.06	0.09
	0.92														0.03	0.06
	0.93															0.03
	0.94															
	0.95															

Figure 12-6. Power Factor Correction Capacity Quick Reference Chart
Source: US Motors
Multiply the motor kW by the factor to arrive at the capacity KVAR required.

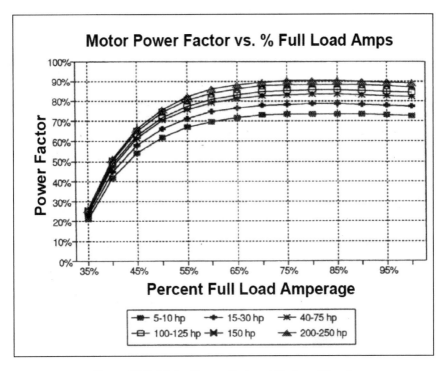

Figure 12-7. Motor Power Factor at Reduced Load
Source: Stum and Koran, "Techniques and Tips for Retro Commissioning Energy Calculations," Presentation at the National Conference on Building Commissioning, 2007. Original Data Source as noted. Original Data Source: BPA, Energy Efficient Elec. Motor Selection Handbook, 1995, p. 39

Equipment	PF	Remarks
Fluorescent Lighting Ballast - Magnetic	0.4-0.6	
Fluorescent Lighting Ballast – Electronic – Normal Ballast	0.8-0.9	
Fluorescent Lighting Ballast – Electronic – High PF Ballast	0.95	
CFL (Compact Fluorescent) - Normal Ballast	0.6	
CFL (Compact Fluorescent) - High PF Ballast	0.95	
HID (High Intensity Discharge) Lighting Ballast, Magnetic	0.4-0.8	
Solenoids, Other Electro-magnets	0.2-0.5	
Induction Heating Equipment	0.6-0.9	
Small "dry" Transformers	0.3-0.9	Reduces with load
Welding – Transformer or Rectifier Type	0.2-0.4	
Welding – Inverter Type	0.9	
Rectifiers	0.8	Reduces with load
Induction Motors 3-100 hp	0.8-0.9	Reduces with load
Small Induction Motors	0.55-0.8	Reduces with load

Figure 12-8. Power Factor for Some Equipment

Chapter 13

Combustion Equipment and Systems

STEAM COST

Example Source: *Energy Management Handbook*, 6th Ed, Turner/Doty.

Example calculation for cost per 1000 lbs of steam.
Assume 200 psig steam, 160 deg F feed water, 82 pct efficiency, fuel cost $2.20 per MMBtu
Enthalpy of steam: 1199 Btu/lb
Enthalpy of feed water: 130 Btu/lb
Heat added per lb of steam: 1199 – 130 = 1069 Btu/lb
Fuel Btu required to make steam: 1069/0.82 = 1304 Btu/lb
Cost of Steam: 1304/1,000,000 * 2.20 *1000 = $2.87 per 1000 lbs of steam

Equipment	Efficiency
Water heater- gravity flue	75-80%
Water heater – forced draft non-condensing	80-85%
Water heater or pool heater - condensing	90-95%
Packaged HVAC Rooftop Unit Furnace	80%
HVAC Furnace, standard	80%
HVAC Furnace, condensing	95%
Boiler, 'tray' type burner, gravity flue	75-80%
Boiler, forced draft non-condensing	80-85%
HW Boiler, condensing **	90-92%

**assumes compatible low temperature distribution temperatures that allow condensing.

Figure 13-1. Combustion Efficiency for Some Equipment
Values are approximate, for new condition equipment.
In general, gravity flues operate with higher excess air and will have lower efficiency than forced draft equipment. Older equipment efficiencies are often less. Higher product temperatures result in lower efficiencies.

278   Commercial Energy Auditing Reference Handbook

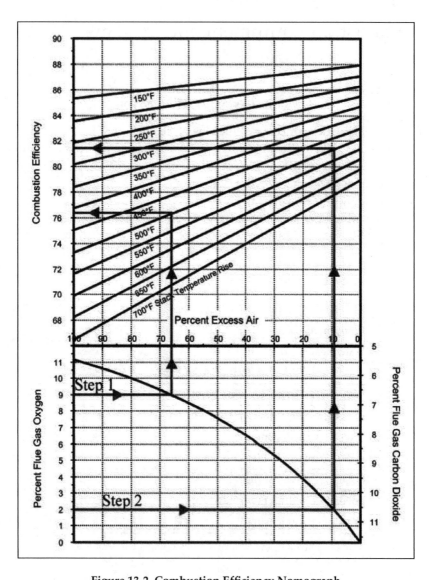

Figure 13-2. Combustion Efficiency Nomograph
Source: "Actions You Can Take to Reduce Heating Costs," Fact Sheet PNNL-SA-43825, Jan 2005, US DOE Office of Energy Efficiency and Renewable Energy.

Combustion Equipment and Systems

Step 1 and Step 2 are before and after examples. Start with either pct O_2, pct CO_2 or pct excess air, and intersect the lower curve, then move directly up on the chart to intersect the flue gas temperature line; read combustion efficiency to the left.

BOILER HEATING OUTPUT WHEN ONLY HEATING SURFACE AREA IS KNOWN

Approximate ratio is 5 SF of heating surface area per boiler hp, and 1 boiler hp = 33,500 Btuh, so Mbh Output = (Heating SF/5)*33.5.

Fuel	Minimum Temp, deg F
Oil Fuel, >2.5%S	390
Oil Fuel, <1.0%S	330
Bituminous Coal, >3.5%S	290
Bituminous Coal, <1.5%S	230
Pulverized Anthracite	220
Natural Gas	220

Figure 13-3. Minimum stack Gas Temperatures to Avoid Corrosion Problems

Source: *Handbook of Energy Engineering*, 5th Ed, Thumann/Mehta

BOILER STANDBY HEAT LOSS (BOILER SKIN LOSS)

About 1.5-2% of full load output.

Source: APOGEE Interactive, Inc.

These are for packaged boilers found in commercial buildings. For larger utility boilers the losses are less, in the range of 0.5% to 1% depending upon size. This is due to the larger units having a more favorable ratio of internal volume to surface area.

ESTIMATED LOSSES FROM BOILER SHORT CYCLING

Thermal losses are typically 1.5-2% of full load output. At reduced loads, the skin losses become a higher fraction of the total load on the boiler, effectively reducing its thermal efficiency. This is especially pronounced when short cycling a large boiler in mild weather or when using a winter heating boiler side-arm heater for summer domestic water heating. E.g. 2-2.7% at 75% load, 3-4% at 50% load, 6-8% at 25% load.

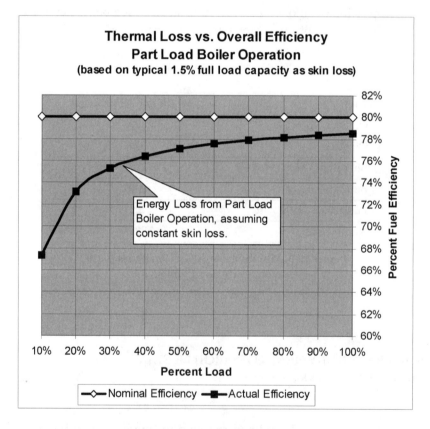

Figure 13-4. Thermal Efficiency Reduction at
Part Load Boiler Operation

(*Continued*)

Combustion Equipment and Systems

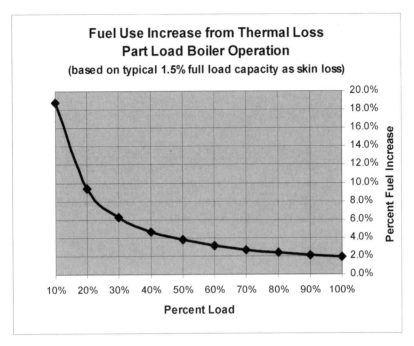

Figure 13-4. Thermal Efficiency Reduction at
Part Load Boiler Operation (*Continued*)

ESTIMATED SAVINGS FROM STACK DAMPERS AND BOILER ISOLATION VALVES

When hot water is pumped through a boiler that is off, the boiler heat exchanger acts in reverse and the boiler loses heat. In some cases this heat loss can be pronounced. How effective the 'heater' is depends on several factors including the amount of surface area for conduction to the room, the temperature of the room, and how well the chimney effect will be to sweep the heat out of the boiler through the flue.

The table values below are based in part on readings from a process control screen shot of multiple boilers operating with flow through both "on" and "off" boilers. The observed savings were de-rated heavily to be conservative, but are still based on limited data. An attempt was made to subjectively proportion the savings to casing loss and flue loss in order to associate approximate savings to measures where isolation valves and stack dampers would be added. The table suggests that isolation valves are a good bet in most cases, however stack dampers, in some applications, provide little benefit.

Note: In the case of a single pump delivering only half of the design water flow through a firing boiler; because the other half is going through the 'off' boiler, there are additional savings from adding valves. The reduced water flow will result in higher temperature rise through the running boiler (mixing with the other one) and notably higher flue gas temperature as a result. Rule of thumb value is 1pct penalty for each 40 degree rise, but this can be higher for certain boilers. See **Chapter 8** — Building Operations and Maintenance **"Approximate Efficiency Improvements from Cleaning Fouled Heat Exchangers (Natural Gas)"** for additional detail.

Approximate Savings for a Two Boiler System
By Adding Isolation Valves and/or Stack Dampers
Savings apply only to the hours the boiler is off and hot - these are not annual savings

Approximate Savings During Hot Idle Times	Draft Damper Savings - Alone, with Hot Water Flowing Through the 'Off' Boiler		
13%	Natural draft (collar)	Flue Common with other Boilers	Constant temp, no reset
7%	Natural draft (collar)	Independent Flue	Constant temp, no reset
3%	Natural draft (collar)	Draft Diverter	Constant temp, no reset
7%	Natural draft (collar)	Flue Common with other Boilers	Boiler Reset
3%	Natural draft (collar)	Independent Flue	Boiler Reset
2%	Natural draft (collar)	Draft Diverter	Boiler Reset
3%	Natural draft (collar)	Flue Common with other Boilers	Domestic Water Heater or Pool Heater
2%	Natural draft (collar)	Independent Flue	Domestic Water Heater or Pool Heater
1%	Natural draft (collar)	Draft Diverter	Domestic Water Heater or Pool Heater
3%	Forced draft	Flue Common with other Boilers	Constant temp, no reset
2%	Forced draft	Independent Flue	Constant temp, no reset
1%	Forced draft	Draft Diverter	Constant temp, no reset
2%	Forced draft	Flue Common with other Boilers	Boiler Reset
1%	Forced draft	Independent Flue	Boiler Reset
0%	Forced draft	Draft Diverter	Boiler Reset
1%	Forced draft	Flue Common with other Boilers	Domestic Water Heater or Pool Heater
0%	Forced draft	Independent Flue	Domestic Water Heater or Pool Heater
0%	Forced draft	Draft Diverter	Domestic Water Heater or Pool Heater
	Isolation Valve Savings - Alone		
9%			Constant temp, no reset
5%			Boiler reset

Sample Calculation to determine savings.
First, estimate the average fraction of hot idle time of both boilers combined.
If, during the heating season, the lead boiler is idle 40% of the time and the lag boiler is idle 80% of the time, then the average combined idle time is (80+40)/2 = 60% idle time. This example will be for isolation valves only, at 9%. For this example, assume gas use is for heating and domestic hot water only. Estimate the fraction of gas use for domestic hot water; this example will use 10%, leaving 90% of the total gas use for the boilers.
If annual gas use is 20,000 therms, then the savings would be estimated at
20,000 * 90% * 60% * 9% = 972 therms per year, or 4% reduction in boiler energy use.

Figure 13-5. Approximate Savings from Boiler Isolation Valves and/or Stack Dampers

Heat Generation Opportunities	Potential Savings
Control air-to-fuel ratio at all loads	5 to 25%
Preheat combustion air	15 to 30%
Oxygen enriched combustion air	5 to 25%
Heat Transfer Opportunities	
Improve Heat Transfer with Advanced Burners and Controls	5 to 10%
Improving Heat Transfer within a Furnace	5 to 10%
Heat Containment Opportunities	
Reduce wall heat losses	2 to 5%
Furnace pressure control	5 to 10%
Maintain door and tube seals	up to 5%
Reduce cooling of internal parts	up to 5%
Reduce radiation heat losses	up to 5%
Heat Recovery Opportunities	
Combustion air preheating	10 to 30%
Fluid or load preheating	5 to 20%
Heat cascading	5 to 20%
Fluid heating or steam generation	5 to 20%
Absorption cooling	5 to 20%
Enabling Technologies Opportunities	
Install high turndown combustion systems	5 to 10%
Programmed heating temperature setting for part load operation	5 to 10%
Monitoring and control of exhaust gas oxygen and unburned hydrocarbon and carbon monoxide emissions	2 to 15%
Furnace pressure control	5 to 10%
Correct location of sensors	5 to 10%

Figure 13-6. Estimated Savings of Process Heating Equipment Improvements

Source: "Improving Process Heating System Performance," Industrial Technologies Program (ITP), 2006, US DOE Office of Energy Efficiency and Renewable Energy. Savings are approximate.

ESTIMATED SAVINGS OF STEAM SYSTEM IMPROVEMENTS

Reduce Steam Pressure
Source of Data: "Reducing Steam Header Pressure Provides Attractive Operating Cost Savings," Office of Industrial Technology (OIT), 2000, US DOE Office of Energy Efficiency and Renewable Energy.

One customer reported an 8% energy reduction by reducing steam header pressure from 125psig to 100 psig.

Trap Orifice Diameter, inches	Steam Loss, lbs. per hour			
	15 psig	100 psig	150 psig	300 psig
1/32	0.85	3.3	4.8	—
1/16	3.4	13.2	18.9	36.2
1/8	13.7	52.8	75.8	145
3/16	30.7	119	170	326
1/4	54.7	211	303	579
3/8	123	475	682	1303

Figure 13-7. Steam Leak Discharge Rate Table

Source of Data: "Steam Tip Sheet #1," Industrial Technologies Program (ITP), 2006, US DOE Office of Energy Efficiency and Renewable Energy.
Original data from the Boiler Efficiency Institute. Steam is discharging to atmosphere through a re-entrant orifice with a coefficient of discharge equal to 0.72.

Combustion Equipment and Systems

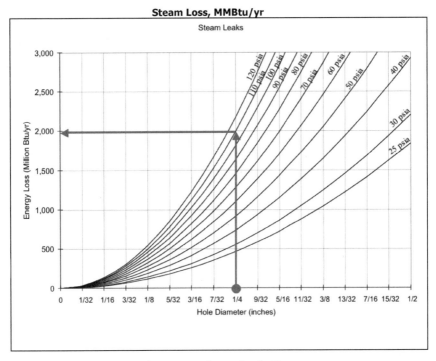

Figure 13-8. Steam Leak Chart
Source: "Actions You Can Take to Reduce Heating Costs," Fact Sheet PNNL-SA-43825, Jan 2005, US DOE Office of Energy Efficiency and Renewable Energy.

Distribution Line Diameter, inches	15 psig	150 psig	300 psig	600 psig
1	140	285	375	495
2	235	480	630	840
4	415	850	1120	1500
8	740	1540	2030	2725
12	1055	2200	2910	3920

Figure 13-9. Heat Loss for Un-insulated Steam Line
Source of Data: "Steam Tip Sheet #2," Industrial Technologies Program (ITP), 2006, US DOE Office of Energy Efficiency and Renewable Energy.
Based on horizontal steel pipe, 75 deg F ambient air, no wind velocity, and 8760 operating hours per year.
Units are MMBtu/yr per 100 ft of uninsulated line

Initial Stack Gas Temperature, deg F	25 MMBtu/hr Boiler Thermal Output	50 MMBtu/hr Boiler Thermal Output	100 MMBtu/hr Boiler Thermal Output	200 MMBtu/hr Boiler Thermal Output
400	1.3	2.6	5.3	10.6
500	2.3	4.6	9.2	18.4
600	3.3	6.5	13.0	26.1

Figure 13-10. Recoverable Heat from Flue Gas
Source of Data: "Steam Tip Sheet #3," Dept of Energy, Industrial Technologies Program (ITP), 2006, US DOE Office of Energy Efficiency and Renewable Energy.
Based on natural gas fuel, 15% excess air, and a final stack temperature of 250 deg F.
Units are MMBtu/hr.

Blowdown Rate, % Boiler Feedwater	50 psig	100 psig	150 psig	250 psig	300 psig
2	0.45	0.5	0.55	0.65	0.65
4	0.9	1.0	1.1	1.3	1.3
6	1.3	1.5	1.7	1.9	2.0
8	1.7	2.0	2.2	2.6	2.7
10	2.2	2.5	2.8	3.2	3.3
20	4.4	5.0	5.6	6.4	6.6

Figure 13-11. Recoverable Heat from Boiler Blowdown
Source of Data: "Steam Tip Sheet #10," Industrial Technologies Program (ITP), 2006, US DOE Office of Energy Efficiency and Renewable Energy.
Based on a steam production rate of 100,000 pounds per hour, 60 deg F makeup water, and 90% heat recovery.
Units are MMBtu/hr.

Combustion Equipment and Systems

Operating Temperature, deg F	3 Inch Valve	4 Inch Valve	6 Inch Valve	8 Inch Valve	10 Inch Valve	12 Inch Valve
200	800	1,090	1,560	2,200	2,900	3,300
300	1,710	2,300	3,300	4,800	6,200	7,200
400	2,900	3,400	5,800	8,300	10,800	12,500
500	4,500	6,200	9,000	13,000	16,900	19,700
600	6,700	9,100	13,300	19,200	25,200	29,300

Figure 13-12. Energy Savings from Installing Insulated Valve Covers
Source of Data: "Steam Tip Sheet #17," Industrial Technologies Program (ITP), 2006, US DOE Office of Energy Efficiency and Renewable Energy.
Based on installation of a 1-inch thick insulating pad on an ANSI 150 pound-class flanged valve with an ambient temperature of 65 deg F and zero wind speed.
Units are Btu/hr.

Excess % Air	Excess % Oxygen	200**	300**	400**	500**	600**
9.5	2.0	85.4	83.1	80.8	78.4	76.0
15.0	3.0	85.2	82.8	80.4	77.9	75.4
28.1	5.0	84.7	82.1	79.5	76.7	74.0
44.9	7.0	84.1	81.2	78.2	75.2	72.1
81.6	10.0	82.8	79.3	75.6	71.9	68.2

Figure 13-13. Estimated Savings from Reducing Excess Air
Source of Data: "Steam Tip Sheet #4," Industrial Technologies Program (ITP), 2006, US DOE Office of Energy Efficiency and Renewable Energy.
Assumes complete combustion with no water vapor in the combustion air.
**Flue Gas Temperature Minus Combustion Air Temperature, deg F
Units are % combustion efficiency (natural gas).

	Per kW	Per HP
Diesel Reciprocating Engine	0.08 gph/kW	0.06 gph/HP
Nat Gas Reciprocating Engine (1000 Btu/CF)	13 cfh/kW	9 cfh/kW

Figure 13-14. Generators: Approximate Specific Fuel Consumption
500 kW and larger.
Source: Mfg Literature
Note: Smaller motors may be 30% less efficient

Air Preheaters	Transfer energy from stack gases to incoming combustion air	2.5 % increase for each 100 degF decrease in stack gas temperature
Economizers	Transfer energy from stack gases to incoming feed water	2.5 % increase for each 100 degF decrease in stack gas temperature (1 % increase for each 10 degF increase in feed water temperature)
Fire tube turbulators	Increases turbulence in the secondary passes of fire tube units thereby increasing efficiency by increasing heat transfer	2.5 % increase for each 100 degF decrease in stack gas temperature
Combustion control systems	Regulate the quantity of fuel and air flow	0.25 % increase for each 1% decrease in excess O2, depending upon the stack gas temperature
Oil and Gas burners	Promote flame conditions that result in complete combustion at lower excess air levels	0.25 % increase for each 1% decrease in excess O2, depending upon the stack gas temperature

Figure 13-15. Some Ways to Increase Boiler Efficiency
Source: *Handbook of Energy Engineering*, Fifth Edition, Albert Thumann and D. Paul Mehta, 2001

Chapter 14

Compressed Air

ESTIMATED SAVINGS OF COMPRESSED AIR IMPROVEMENTS

Compressed Air Cost

A general guideline for estimating compressed air cost
@ 10¢ per kWh: $171 per cfm per year continuous use.
Source: Ingersoll-Rand Company Limited

This is based on 4cfm per hp which is the rule of thumb for 100 psi. Higher pressure will require more power.

CFM of Leakage

Pressure (psig)	1/64" hole	1/32" hole	1/16" hole	1/8" hole	1/4" hole	3/8" hole
70	0.29	1.2	4.7	19	74	168
80	0.32	1.3	5.2	21	83	187
90	0.36	1.5	5.7	23	92	207
100	0.40	1.6	6.3	25	101	227
125	0.48	1.9	7.7	31	122	276

Figure 14-1. Compressed Air Leak Chart
Source of Data: "Compressed Air Tip Sheet #3," Industrial Technologies Program (ITP), 2004, US DOE Office of Energy Efficiency and Renewable Energy.
  For well-rounded orifices, values should be multiplied by 0.97.
For sharp orifices, values should be multiplied by 0.61.

Air Compressor Efficiency

Larger machines use screw compressors which are more efficient than reciprocating machines, by 15-20%.
Source: manufacturer's literature.

Savings from Reduced Compressed Air Pressure

1% air compressor energy savings for each 2 psi lowered in compressed air pressure at the source.
Source: E Source Tech Update "Assessing Processes for Compressed Air Efficiency," 1995, cross checked with manufacturer's catalog data.

Savings from Cooler Inlet Air

Each 10 deg F drop in inlet air temperature will save 1.9% energy.
Source: "Investment Grade Compressed Air System Audit, Analysis, and Upgrade in a Pulp and Paper Mill," Paresh S. Parekh, Unicade Inc., 2000

Heat Recovery Potential for Compressed Air

Approximately, 50,000 Btu/hour of energy is available for each 100 cfm of full-load compressor capacity.
Source: "Investment Grade Compressed Air System Audit, Analysis, and Upgrade in a Pulp and Paper Mill," Paresh S. Parekh, Unicade Inc., 2000

Efficiency of Conversion for Compressed Air, at the Source

Approximately 28% efficiency at the source.

Based on the rule of thumb of 50,000 Btuh per 100 cfm, and 4 cfm per hp at 100 psig, and a 90% efficient motor, the power input for 100 cfm would be (100/4/0.90 = 27.7 hp). The equivalent heat from 27.7 hp is 27.7 *0.746 *3413 = 70,526 Btuh.

For this 100 cfm of capacity, the input hp converted to Btu is 70,526 Btu of input heat (electrical equivalent) is compared to 50,000 of air compressor heat that is rejected, thus 20,526 Btu of the total input ended up as energy in the compressed air.

Using rough numbers: Based on rejected heat, the compression process efficiency is approximately (70-50)/70= 28% efficient at the source.

Efficiency of Conversion for Compressed Air, at the Point of Use

Approximately 22% efficiency at the point of use.

Begin with prior rule of thumb of 28% (See "**Efficiency of Conversion for Compressed Air, at the Source**," this section)

Assume a loss of 15% of the system volume, through leaks, purge air, etc.
Source: E Source Tech Update "Assessing Processes for Compressed Air Efficiency," 1995.

Author's note: this text used half of the stated typical value of 20-40%. Assume 10% of the initial pressure is lost from pipe friction and pres-

Compressed Air

sure regulation devices. The point of use compressed air efficiency is then 28/1.1/1.15=22% efficient at the point of use.

Source: Derived from E Source Tech Update "Assessing Processes for Compressed Air Efficiency," 1995.

Regenerative Air Driers (Desiccant Type)

These have a parasitic loss of about 15% of the air supplied.

Source: manufacturer's catalog data.

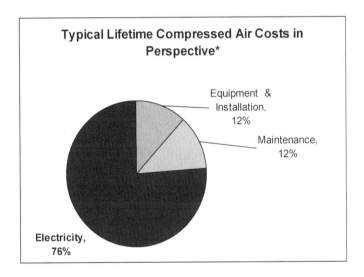

Figure 14-2. Typical Lifetime Costs of Compressed Air
Source of Data: "Compressed Air Tip Sheet #1," Industrial Technologies Program (ITP), 2000, US DOE Office of Energy Efficiency and Renewable Energy.
Assumptions in this example include a 75 hp compressor operated 2 shifts a day, 5 days a week at an aggregate electric rate of $0.05/kWh over 10 years of equipment life.

Chapter 15

Fans and Pump Drives

FAN DRIVE EFFICIENCY COMPARISON

NOTE: Predicted savings require a de-rate to account for maintaining a constant downstream pressure. See **Chapter 9—Quantifying Savings "VAV System Fan Savings Reduction for Maintaining Downstream Pressure"** for more information on this effect and a table of correction factors.

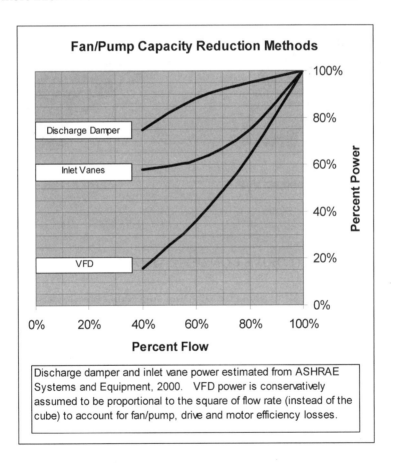

Figure 15-1. Fan Drive Efficiency Comparison Diagram

293

COG BELTS INSTEAD OF STANDARD V-BELTS

Basis of Savings: Cog Belts bending easier, which translates into a more efficient belt drive coupling.

- The notches in the belt inner race allow the repeated bending of the belt around sheaves to occur with less effort. Power and energy consumption are normally reduced by 2% due to higher efficiency drive coupling.
- Special sheaves are normally not required for this measure, making it easy to implement as a maintenance practice.

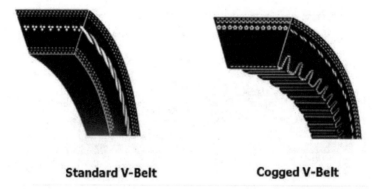

Standard V-Belt   **Cogged V-Belt**

Source of belt photos: Carlisle

VARIABLE SPEED DRIVE CONSIDERATIONS

A successful VFD project is one that succeeds both functionally and financially. Here are some considerations.

The financial hurdle requires identifying the number of hours of operation at various percent of full load, and the efficiency of the VFD compared to the existing drive. The load profile is essential to avoid over-stating savings.

Load Profile
Determine:
- Hours/year from 0 - 60% flow
- Hours/year from 60 - 80% flow

Measured energy use with V-Belt vs. Cogged Belts
Source: "Energy Comparison Study for Standard (Wrapped) Versus Cogged V-Belts," Industrial Technology Institute and Detroit Edison, November, 10, 1995.

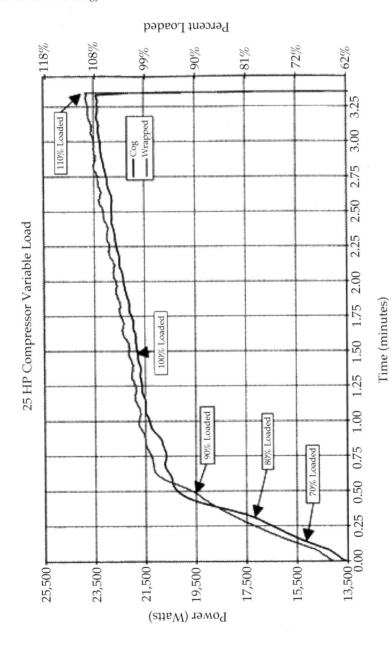

- Hours/year from 80 -100% flow

Assuming the system is not altered other than the drive, the analysis is fairly easy once the load profile is established. The differential efficiency at each category of load, multiplied by the hours at those loads, is it.

VFD Savings: Square Instead of Cube

Using the affinity laws (See **Chapter 21—Formulas and Conversions** "**Affinity Laws**") suggests that savings diminishes proportionally as the cube of the speed change. This is true for the fluid power, but does not hold true all the way up the food chain to the electrical input. The reasons for this are that motor, fan and pump efficiencies fall off at reduced speeds (for various reasons), and consume some of the savings. One way is to assume power reductions are proportional to the square of flow rate—rather than the cube.

Here is a modified affinity law to consider, which will yield more realistic results.

Affinity Law, Modified for VFD Savings:
$HP2 = HP1 \times (N2/N1)^{2.0}$ ← use square instead of cube

Savings at Very Low Speeds

Predictions of savings at motor speeds less than 40% may not be realistic. In addition to the motor and drive losses, the process itself may behave differently at low loads from things like laminar flow in coils or overlapping heating and cooling.

For example, trend VAV air flow percent throughout the year. Compare actual minimum air flows of the fan to the weighted average of the VAV box minimums. If the fan does not achieve the intended minimums, ask "why is that?" More importantly use caution in accepting computer predictions of very low flows for variable flow water/air systems.

Savings When Controlling to a Constant Downstream Pressure

Predicted savings require a de-rate to account for maintaining a constant downstream pressure.

For more information on this effect and a table of correction factors, See **Chapter 9—Quantifying Savings**, "**VAV System Fan Savings Reduction for Maintaining Downstream Pressure.**"

Chapter 16

Lighting

GENERAL

- About 12% of the cost of light [in its lifetime] is from the hardware and maintenance. The other 88% is from energy use.
  Source: "Found Money: A CFO's Guide to the ROI of Lighting", Energy & Power Management, July 2005

- Because the ballast mostly determines how many watts are used, ballast choice is critical to the energy efficiency success of a project.
  Source: "FEMP Lights" Training Material, Federal Energy Management Program, 2005. Original source of data Heschong Mahone Group, Inc.

- During summer, lighting savings have a double-dip savings effect for air conditioned spaces, since the excess wattage becomes heat which becomes load on the A/C system. Dependency upon the number of cooling hours, the A/C savings from lighting retrofits may be significant or may be ignored.

- Moonlight on a cloudless night on a white beach is about 1 foot candle.

- During winter, the waste heat from lights can be a supplement to the building heating system, but is normally not considered. However, lighting retrofits that reduce energy use may impact the overall building heating system capacity and must be checked.

- Indirect Lighting is typically 15% less efficient than direct systems, because the light must first bounce off the ceiling.
  Source: "FEMP Lights" Training Material, Federal Energy Management Program, 2005. Original source of data Heschong Mahone Group, Inc.

- Costs of turning lights off and on.
  Source: "FEMP Lights" Training Material, Federal Energy Management Program, 2005. Original source of data Heschong Mahone Group, Inc.
  — Fluorescent: The economic break-even point is typically between 5 and 15 minutes between switching.

— HID: The economic break-even point is around one hour between switching.

LIGHTING TERMS

Light output of fixtures and bulbs is rated in **lumens**, measured at the source.

Foot-candles (FC) is the standard measure of illuminance. Units are lumens per square foot, usually measured at the work surface.

Lux is the metric unit for illuminance, in lumens per square meter. To convert FC to Lux, multiply foot-candles by 10.76.

Higher **color temperatures** mean more white/blue appearance. Units are degrees K (Kelvin). A 4100K light source is seen as blue or white, and a 3000K light source is more yellow and warm color.

CRI is a 0-100 scale indicating how perceived colors match actual colors. The higher the number, the closer the match. This is an industry standard test using (8) different colors and a reference light source. Incandescent lighting has a CRI of 100.

Efficacy is a term used for evaluating lighting sources and is in units of **lumens per watt**.

Ballast factors can be selected to alter light output and energy use; in effect tuning the lighting system. The ballast factor is the ratio of a light output using the selected ballast, compared to the rated light output. General purpose ballasts have a ballast factor less than one; special ballasts may have a ballast factor equal to or greater than one.

Example of controlling energy use with ballast factor: A fluorescent lighting retrofit is expected to produce 55 foot candles at the work surface compared to 40 foot candles existing that is acceptable to the customer. In addition to the savings of more efficient bulbs and ballasts, selecting a ballast factor of 0.70 instead of 0.87 will reduce lighting power and light by a factor of $0.70/0.87 = 0.80$, and reduce foot candles from 55 to 41, providing an additional 20% energy savings.

DIMMING

Incandescent: Easy, just a rheostat.

Fluorescent: The reduction in power is not as great as the reduction in light output, therefore efficiency declines somewhat with dim-

Lighting 299

ming. Requires special ballast. Mixed success with Compact Fluorescents (CFL).

HID: Problematic, not common. However, bi-level switching does work well.

LIGHT COLORED SURFACES

Reflective (light) colored surfaces increase effectiveness of the lighting since less of it is absorbed. This includes floors, furniture, walls, and ceilings.

It can take up to 40% more light to illuminate a dark room than a light room with a direct lighting system.
Source: "FEMP Lights" Training Material, Federal Energy Management Program, 2005. Original source of data Heschong Mahone Group, Inc.

Reflectance Parameters for Picking Interior Surfaces and Colors

Min 80% reflective   Ceiling
Min 50% reflective   Walls
Min 25% reflective   Floor and furniture

Color/tone	Reflectivity
Bright White	80-90 %
Light Tint	60-80 %
Pastels	40-60 %
Bright Colors	20-40 %
Deep Colors	10-30 %
Dark Colors	5-10 %

Figure 16-1. Reflective Values of Common Colors
Source: "FEMP Lights" Training Material, Federal Energy Management Program, 2005. Original source of data Heschong Mahone Group, Inc.
   Photographic "middle gray" and Caucasian flesh tones are about 20% reflective.
   A "true" black absorbs almost 100% of the light that strikes it 5-15% reflectance (85-95% absorption)—dark furniture, carpet.

GENERAL LIGHTING INFORMATION

Type	Efficacy (lumens per watt)	Typical Life (hrs)	Color Rendition	Re-light time
Incandescent	10	750	Excellent	Immediate
Tungsten Halogen	15	3,000 hrs	Excellent	Immediate
Mercury Vapor	50-60	16,000 to 24,000	Poor to Fair	3-10 minutes
Compact Fluorescent (CFL)	50-75	10,000-20,000	Good to Excellent	Immediate
Fluorescent	75-90	7,500-24,000	Good to Excellent	Immediate
Metal Halide	80 - 110	9,000-20,000	Fair for regular, Excellent for Ceramic	3 to 15 minutes
Super T-8 Fluorescent	90 to 100	24,000 to 30,000	Excellent	Immediate
High Pressure Sodium	65 to 125	24,000 to 30,000	Fair	1 minute
Low Pressure Sodium	Up to 180	18,000	Poor	Immediate

Figure 16-2. Lighting Technology Properties
Common technologies
Arranged by Efficacy (lumens per watt)
Note: lumens per watt is for the bulb only
Source: Osram Sylvania 2004 Lamp and Ballast Product Catalog

Lighting

Sector	Percent
Offices	40%
Health Care	30%
Hotel/Hospitality	30%
Residential	25%
Industrial	20% (highly variable)
Retail	55%
Warehouse	40%

Figure 16-3. Lighting Energy Use, Pct of Total Electric, by Building Type
Source: "FEMP Lights" Training Material, Federal Energy Management Program, 2005. Original source of data Heschong Mahone Group, Inc.

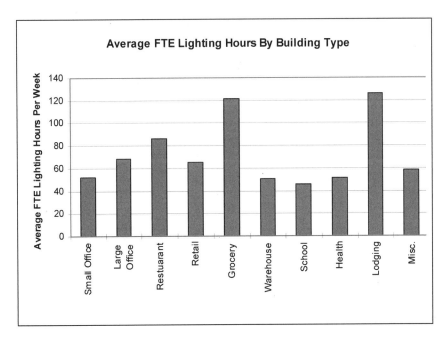

Figure 16-4. Average Lighting Hours by Building Type
FTE = Full Time Equivalent.
Source: "FEMP Lights" Training Material, Federal Energy Management Program, 2005. Original source of data Heschong Mahone Group, Inc.

Business	Ambient FC	Task FC	Remarks
Office Bldg	20-50	50 Reading 30 Computer eyboards	10-20 Lobbies
Department Store	50-100	200-50 Cashier	Discount stores generally higher light levels
Jewelry	30-75	200-500 Manufacture Cashier 20-50	150-500 Accent lighting
Merchandising	30-75	20-50 Cashier	150-500 Accent lighting
Furniture showroom	10-30 appraisal	30-100 on furniture	150-500 Accent lighting
Schools - classroom	50 Reading	30 Computer keyboards	50-100 Labs
Restaurant dining sitting area	5-10 Dining	50-100 Kitchen	10-20 Cleaning
Hotel guest room	10-20	20-50 Bathroom	10-20 Lobbies
Grocery Store	100 (vertical) high activity 75 Medium activity 30 Low activity	20-50 Cashier	
Warehouse	10 Inactive	10-50	bulky to small items
Church	50-75	100-150	
Kitchen	50-100	50-100	
Residential	5-10	50-100	
Parking Garage (enclosed)	50 entrance 5 General Parking	---	
Parking Lot (exterior)	0.5-2	---	

Figure 16-5. Typical Recommended Lighting Levels

Source: Stephen W. Leinweber, L.C., C.L.E.P.
These values are only intended to help identify areas that are substantially over-lit. For example for the stated range of 30-50 FC, a measured value of 51 is not cause for alarm, but a measured value of 75 suggests de-lamping may be viable. This is not intended as a substitute for a lighting designer's judgment—and the customer may want or need the additional light.

Lighting 303

LIGHTING OPPORTUNITIES
Source: "TEMP Lights" Training Material, Federral Energy Management Program, 2005. Original source of data Heschong Mahone Group, Inc.

Definite
- Four lamp fluorescent troffers using T-12 lamps and standard magnetic ballasts
- Mercury vapor lighting of almost all kinds
- Incandescent down-lights in public spaces, that are not dimmed for functional or aesthetic purposes
- Incandescent EXIT signs
- Spaces that are over-lit, especially if removing lamps or fixtures is appropriate
- Retail space employing R-40 or ER-40 track lighting

Maybe
- 2 + 3 lamp troffers, T-12 lamps and magnetic ballasts
- Industrial fluorescent lighting
- Private offices without occupancy sensors
- School and retail T-12 Fluorescent lighting
- Fluorescent EXIT signs
- Any location with short life lamps and high maintenance costs
- Any location being renovated for other reasons
- Locations appropriately converted to a task/ambient lighting system

Slim Chance
- Lighting with T-12 lamps and magnetic ballasts with operating hours limited by occupancy sensors or by a building automation system
- Incandescent lighting with dimming
- Any lighting in service spaces and living quarters where operating hours may be very short
- Areas where day-lighting suggests installations of photo controls
- Open offices or public rooms without occupancy sensors or other controls

Energy Cost Rate
- Buildings which enjoy very low utility rates are mediocre retrofit candidates, because payback will be longer and Savings to Investment

ratio (SIR) will be lower. A low rate suggests caution in proceeding further.

Hours of Operation
- The longer the hours of operation, the more attractive a retrofit will be.
- More than 5000 hours per year—good chance
- Less than 2500 hours per year—slim chance

OCCUPANCY SENSOR ENERGY SAVINGS
Source: Watt Stopper/Legrand, Technical Bulletin #151, 2002.

Testing was part of EPA's Green Lights Program (2001) and included a total of 158 rooms falling into 5 occupancy types: 42 restrooms, 37 private offices, 35 classrooms, 33 conference rooms and 11 break rooms.

Energy waste is the total waste from lighting on while unoccupied, and energy savings is the amount recouped by the use of occupancy sensors, with a 20 minute time delay.

Application	Energy Savings (20-min. time out)	Energy Waste
Break room	17%	39%
Classroom	52%	63%
Conference room	39%	57%
Private office	28%	45%
Restroom	47%	68%

Figure 16-6. Occupancy Sensor Energy Savings
Source: Watt Stopper/Legrand, Technical Bulletin #151, 2002.

Chapter 17

Envelope Information

In many commercial building, envelope losses are minor compared to energy use by activities inside the building. But there are exceptions where envelope loads are a substantial part of total energy use, such as:
- Buildings with very light internal activities
- Residential use (hotels, motels, dormitories)
- All-glass buildings

BLC HEAT LOSS METHOD

Annual heating energy use, for buildings whose heating use is dominated by envelope losses, can be estimated using the building load coefficient (BLC) method with reasonable accuracy.
Source: *Energy Management Handbook* 6th Ed, Turner/Doty.

BLC Equation

$$E = BLC * DD * 24$$

Where:
E=energy output, Btu (to arrive at input, divide by appropriate efficiency)
DD=Heating Degree Days
24 = conversion from Degree Days to Degree-Hours
BLC=Building Load Coefficient
  BLC= (sum U*A) + (sum F*P) +0.018 * (Q_{infilt}) +1.1 *(Q_{vent})
  (sum U*A)=summation of (U_{value}*Area) for components (walls, windows, roof)
  (sum F*P)=summation of (F-factor * Perimeter LF) for slab on grade.
    F- (un-insulated slab) = 0.73 Btu/hr-LF-deg F
    F- (R-10 under slab) = 0.54 Btu/hr-LF-deg F
  Q_{infilt} = infiltration, cubic feet per *hour*
  Q_{vent} = infiltration, cubic feet per *minute*

Input Energy

The BLC calculation yields the output energy for the envelope. To calculate the input, divide the output by the heating/cooling equipment efficiency.

BLC and Cooling Loads

Note: Use of the BLC method for cooling loads will understate actual loads, since this method does not include internal loads, dehumidification loads, or solar loads that add substantially to actual cooling loads.

R-VALUE REDUCTION FROM STUD WALLS

Size of members	Framing	Insulation R-value	Metal Stud Correction Factor	Wood Stud Correction Factor
2x4	16 in. O.C.	R-11	0.50	0.76
2x4	24 in. O.C.	R-11	0.60	0.80
2x6	16 in. O.C.	R-19	0.40	0.60
2x6	24 in. O.C.	R-19	0.45	0.66

Figure 17-1. Parallel Path Correction Factors
(reduction in overall R-Value of the wall)
Source: *Energy Management Handbook* 6th Ed, Turner/Doty

Size of members	Framing	Metal stud example before and after R-value	Metal stud example before and after R-value
2x4	16 in. O.C.	R-11/R-5.5	R-11/R-8.4
2x4	24 in. O.C.	R-11/R-6.6	R-11/R-8.8
2x6	16 in. O.C.	R-19/R-7.6	R-19/R-11.4
2x6	24 in. O.C.	R-19/R-8.6	R-19/R-12.5

Figure 17-2. Effective Insulation De-Rate Effect from Stud Walls
Source: Author Calculations derived from *Energy Management Handbook* 6th Ed, Turner/Doty

Envelope Information

GLAZING PROPERTIES

Thermal
- Glass by itself has meager insulating properties. Increased thermal insulation is achieved by adding layers and pockets of trapped air or inert gas. Common center of glass U-values:

U-Value	# of Panes
1.1	single pane
0.5	standard double pane
0.3	standard triple pane (note diminishing return)

- The addition of coatings and different gases influences the overall U-value, as does the frame construction (metal, wood, vinyl, with-without thermal breaks).
- The effect of the frame on thermal performance should not be overlooked. A plain double pane window with a wood or vinyl frame has similar performance to that of Low-E coating (hard coating) with Argon gas fill and a metal frame.
  Source: *Energy Management Handbook*, 6th ed, Turner/Doty.

High Performance Glazing
- High performance glazing systems can sometimes eliminate perimeter heating systems with equipment savings that help pay for the glazing. For example, glazing systems are available and can provide an inside surface temperature of 55 deg F or higher at (–10) deg F outside temperature. The high R-values are achieved by multiple layers, each one adding a trapped air (or gas) space. These can be multi-layered glazing units or "suspended film" units.

Solar Shading
- Summer maximum cooling loads are often dictated by solar load, so reducing this directly reduces equipment size and cost. Also, occupants in the direct path of incoming sunlight will feel too warm despite the surrounding air temperature, and will lower the thermostat to compensate, adding further energy consumption.

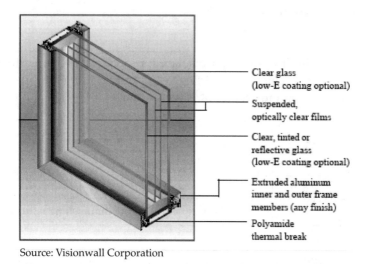

Source: Visionwall Corporation

Figure 17-3.
The above cutaway diagram is an example of a proprietary high performance glass unit manufactured by Visionwall Corporation which achieves a combined glass/frame insulating value in excess of R-7. The increased R-value is achieved by multiple layers of glass and/or film. Note the thermal break in the frame as well.

- Glass performance is rated by the 'shading coefficient' (SC) and 'solar heat gain factor' (SHGF). Common values of modified glass:

SHGF	Treatment
1.0	100% of light passes through (theoretical)
0.9	common clear glass
0.6	light tint
0.4	heavy tint
0.25	Low E coating

- For Low-E coatings, the soft coating is superior in performance, but is subject to abrasion. For multi-layered glazing, soft coating on an internal and protected layer is effective.
- Specially designed shading elements can sometimes be used to

allow solar infiltration in cold months to reduce heating.
- Shading methods vary, but they share a common goal which is to keep the sun's heat out of the building. Interior shades provide some relief to occupants, but much of the heat ends up as air conditioning load since it is already inside. For this reason, exterior shades are more effective.

- Modern coatings can repel much of the solar heat gain that otherwise comes through the glass and heats the interior contents. Unlike old heavy tints, modern high performance coatings can repel more than 50% of solar heat with a minimal amount of visible light loss.
- Silk screen shaded glass with dot patterns provide shading. Automobile windshields, near the edges, are a ready example of this technology. The fabric is embedded in the glass by the manufacturer.

- Solar films applied to the inside surface of existing glass can be cost effective in some cases. A good glazing will reduce solar load by 50% with 75% visible light transmittance, and will not become a heat sink for the reflected energy. These do not have to look like mirrored sunglasses. Spectrally selective coatings target the heat-producing wavelengths to repel. Design life is less than new windows. Films usually require re-application about every 10 years.
- Exterior screening can be applied to certain existing glass or to skylights. These screens look like screen door material, but come in different "percent free area" patterns. A "30% FA" screen pattern blocks 70% of the light, and allows 30% through. These screens are aggressive at providing shade while allowing some visible light to come through, but are not transparent. Attachment and support of these screens is a design challenge.

INFILTRATION

- This is unintended outside air coming into the building due to differential pressures (mechanical, wind, and stack effect) and openings in the envelope. Energy implication is the heating and cooling energy needed to temper it once inside. In humid climates, the load of dehumidifying the extra outside air is added. Units of infiltration can include CFM, CFM/SF of envelope wall, or air changes per hour (ACH).
- When infiltration is excessive, there are usually complaints of cold drafts in winter. The best time to locate points of infiltration is during cold weather, by the use of an infrared thermometer around various exterior points in the building. Return air plenums are especially problematic if not sealed, since they operate in a slight negative pressure anyway. In extreme cases, frozen pipes in return plenums are the result of infiltration.
- Calculating infiltration is difficult and even the best formulas available rely on subjective data. A widely used relationship, called the crack method, is published in the ASHRAE Fundamentals Handbook and includes a variety of tables and factors, and is not repeated here. The only way to know for sure is by leakage testing which is seldom done due to complexity and cost.
- Infiltration values can vary by as much as a factor of 10.
- A handy rule of thumb for commercial buildings for overall infiltration levels (cfm per SF of wall area) is:

CFM/SF of Wall Area	Construction Quality
0.10	Tight
0.30	Average
0.60	Leaky

Chapter 18

Water and Domestic Water Heating

WATER CONSUMPTION FOR WATER-COOLED EQUIPMENT

1 ton-hour = 1.8 gallons of water
or
1 ton-hour = 15 lbs of water
1 ton-hour = 12,000 Btu output, but the rejected heat has the heat of compression in it, so there is more heat to reject than 12,000 Btu.

Factors for heat of compression at normal condensing temperatures are around 1.25, so 12,000 * 1.25 = 15,000.

One lb of water requires about 1000 Btu to evaporate, hence 1 ton-hr = 15 lbs of evaporated water to cool the condenser.

For actual water used, include blow down. The amount of blow down depends on the cycles of concentration which in turn depends on initial water hardness and type of water treatment used.

Cycles of Concentration	Blow down flow, percent added to evaporation
3	50%
4	33%
5	25%
6	20%
7	17%
8	14%
9	13%
10	11%

Blow down (gpm) = $E/(Cycles - 1)$
Where:
E = evaporation rate, gpm
Cycles = cycles of concentration
(4-7 cycles is common)

Figure 18-1. Blowdown vs. Cycles of Concentration (Cooling Tower)

PLUMBING WATER POINTS OF USE

Beginning at the point of use is recommended. For example, replacing a water heater with a more efficient one will save energy, but the new heater could be a smaller heater if there was less use. The savings in incremental equipment size can pay for the point of use changes (faucets, shower heads, etc.)

Fixture	Flow
Shower (old)	Up to 5.0 gpm
Shower (standard)	2.5 gpm
Shower (ultra low flow)	1-1.5 gpm
Lavatory faucet (old)	2.0-3.0 gpm
Lavatory faucet (low flow)	0.5-1.0 gpm
Urinal (old)	3.5 gpf
Urinal (standard)	1.0 gpf
Urinal (low flow)	0.5 or waterless
Water closet (old)	3.5-5.0 gpf
Water closet (standard)	1.6 gpf
Water closet (hi/low flow)	1.0/1.6 gpf

Figure 18-2. Plumbing Fixture Water Flow Rates

DOMESTIC WATER HEATERS

Gas or Electric

In almost all locations, natural gas is cheaper to use for heating purposes.

Gas heating efficiencies vary. Plain gas heaters have a nominal

Water and Domestic Water Heating 313

80%e but quickly decline because the heat transfer surfaces are not designed to be cleaned and thus are never cleaned. 70-75% is more likely for standard commercial gas heaters.

Condensing gas heaters have excellent potential for efficiency since the water temperatures are generally lower than for heating water use, allowing the condensation to occur. 95% efficiency is easily obtainable through condensing for 120 deg F water.

Heat Pump Water Heaters

When gas is not available, or when there is a coincident cooling demand that is always there, heat pump water heaters can be effective. A COP of around 2.0 is expected at 140 degrees condensing, which means it costs half as much as electric resistance.

In locations where ambient temperatures get cold in winter season, the heat pump water heater would lose efficiency, since it needs a source of heat. Savings will be reduced during times when the heat pump water heater absorbs heat from an environment that is being heated by another system.

A manufacturer's rating of "percent standby loss" is no longer in use, but may still be found on older labeled heaters. For example, a water heater with a 360,000 Btuh heater rated for 3.33% standby loss per hour with a 60 gallon tank would be expected to lose 1200 Btuh per hour. Since this same 60 gallon tank could come with a variety of burners, the percentages also vary for different units using the same tank, and this statistical unit can be confusing.

DOMESTIC WATER HEATER STANDBY LOSSES

Storage Tanks for Domestic Hot Water Heaters

Storage tanks are used as thermal flywheels to accommodate temporary high loads that are in excess of average loads, thereby allowing a smaller burner or heater. The designers use a combination of heating capacity (gallons of hot water recovery rate) and tank size to meet the customer usage requirements. The tank has a standby loss from its surface area.

Any un-insulated tanks or bare sections can be justified by insulating them. Also, any un-insulated valves and piping reduces standby loss.

Estimating these losses can be set approximately equal to the heat loss of a cylinder or pipe of equal diameter and length, although this understates losses since there will be bare fittings and they will short circuit the insulation barrier. Loss is proportional to the differential temperature between inside (fluid) and outside (air), so lowering water temperature reduces standby loss.

One rule of thumb for tank loss is **6.5 Btuh/SF** for an 80 degree differential inside to outside.

Source: Table 404.1—Minimum Performance of Water Heating Equipment, DOE, 10 CFR Ch. II, 2005.

Domestic Hot Water Recirculation

Pumps and a return line are used to reduce the waiting time from opening a faucet to hot water delivery. Pump energy is small and the recirculation line is small. The benefit of a hot water recirculation ECM lies in the thermal losses.

With the pump on, the loss in Btu/h is continuous. With the pump off, the loss continues until the recirculation line is equal to ambient temperature, then stops, so turning off the circulating pump when not needed reduces the standby loss. The two common methods used to control the pump are an aquastat and timer.

An <u>aquastat</u> stops the pump when the water returning from the system is sufficiently high. Once it cools off, it starts again. Some users complain that this method short cycles their pumps.

A <u>timer</u> stops the pump during unoccupied hours, but lets it run continuously during occupied times.

Refer to tables in the Appendix "Heat Loss from Un-Insulated Hot Piping and Surfaces" and "Heat Loss from Insulated Piping" for heat losses of piping to help quantify potential savings.

Instantaneous Water Heaters

By generating hot water at the point of use, distribution losses for domestic hot water can be eliminated. In the case of a small point of use that is a long ways away, this is very advantageous. A point of use water heater applied at a central location addresses tank losses but does nothing to address distribution piping losses which are usually the larger of the two. Also, this ECM eliminates the thermal flywheel design feature of the tank method, so consideration of larger heating capacity and infrastructure to support it become design questions.

Note: Instantaneous electric or gas demand can easily double when point of use heaters are applied compared to storage tank heaters.

Chapter 19

Weather Data

DEGREE DAYS

Units of degree day weather data are: ($^{degF\text{-}days}/_{period}$)

Period is usually months or years. City locations are often compared in tables of heating degree days per year. More detailed weather data list degree days by month.

One heating degree day is one degree below the base temperature consistently for a full day.

One cooling degree day is one degree above the base temperature for a full day.

Calculate average temperature from degree days:

Base temp - ($^{HDD}/_{no.\ of\ days\ in\ the\ period}$)

Base temp + ($^{CDD}/_{no.\ of\ days\ in\ the\ period}$)

Example: if November has 600 degree days (65 degree base), the average temperature would be 65- (600/30) = 45 degrees.

Cautions for Using Degree Days
1. The most common use for degree days is for residential heating, to estimate variations between years. For this reason, the common 'base' number for heating degree days (HDD) is 65 degrees, which indicates that at temperatures below 65 the home will likely begin to need heat. This is seldom a good assumption in commercial buildings, other than for hotel guest rooms and other residential occupancies, or where internal loads are very small and envelope loads dominate.
2. The definition of "degree days" includes the average daily temperature, which is the (daily high—daily low)/2. So, two cities with similar daily average temperatures can have much different weather. For example, if two cities have a high of 50 and a low of 10 they will have an "average" temperature of 30 degrees. But if one city has 16 hours at 10 degrees and one hour at 50 degrees and the other

city has the reverse, their heating requirements are obviously much different. For this reason, using modeled heating data for one city and projecting it to another city (as if the city were picked up and moved) using degree days is risky and not recommended. A better expression of weighted average weather data is "bin" data.

3. Where cooling degree days (CDD) are published, they are usually with the same residential 65 degree base. Since most commercial buildings have significant internal loads with thermal break even temperatures much lower than 65, the use of CDD information will usually underestimate cooling loads. Where CDD50 is available (50 degree F base), commercial loads will be better estimated than CDD65. Note that ASHRAE 90.1 climatic data tables are based on CDD50

For example:

Location	CDD65	CDD50
Atlanta	1,246	5,038
Denver	434	2,732

Source of CDD65: NOAA, 1949-2006 avg
Source of CDD50: ASHRAE 90.1 - 2001

4. Degree days, in any event, are limited to conduction loss/gain and ventilation load loss/gain. Degree days do not provide information to estimate internal loads or solar loads.
5. Weather data software can produce temperature data in terms of degree days in specified base temperatures.

BIN WEATHER DATA

This method uses historical data, usually hourly, to express weather parameters such as dry bulb temperature, wet bulb temperature, etc.

Using dry bulb temperature as an example: for each hour that 70 degrees F is recorded at a weather station, the 70 degF 'bin' is incremented by one (1). So, if the 70 degF bin has a value of 100, this means there were 100 hours recorded at 70 degrees F during that period.

Bin data reports are very useful since they establish the proportions of operational time that will be expected for different ambient weather conditions. Accuracy of calculations is enhanced with such data, since ambient conditions also affect equipment loads and efficiencies, i.e. the corresponding equipment efficiency at 70 degrees (and during the 100 hours) will be unique at that temperature and different at other temperatures.

See the **Appendix** for sample bin weather data for several cities.

Cautions for Using Bin Weather Data

1. While bins indicate the number of hours at a certain temperature, they do not indicate when this occurs.
2. Some of the hours may be transient. For example, 10 bin-hours at 35 degF wet bulb may be the combination of 8 hours all at once, and two days where the wet bulb was only at 35 for one hour as the weather changed from 35 to 40 degF wet bulb. If a process, such as a plate frame heat exchanger, is designed to take advantage of free cooling in dry weather but takes an hour or two to switch operating modes to the free cooling mode, this 'hour' is of no real use and including it overstates savings.
3. Some of the hours may be during unoccupied periods. To correct for this, use a weather data source that allows the user to select appropriate time periods, such as 6am-10pm instead of all 24 hours, for facilities that close at night, weekends, etc. Customizing the bins increases the accuracy of the conclusions drawn for them. Sample bins for five cities are shown in the **Appendix** and illustrate this point, and also to illustrate the differences between climate zones.

WEATHER DATA BY DAYS AND TIMES

This format is used for hourly computer analysis programs. It is also useful in quantifying savings when time-of-day information is available. For many ECM proposals, this method is the quickest.

For example, the cell for January 10am is the average temperature observed in this city for each day in January at that time. For 7 day per week operation, it can be said that the temperature shown for January 10am occurs 30 times (30 days per month).

This form of data representation can be useful for measures that are weather-dependent. For example, if a boiler is left idling all summer currently and a measure is proposed to turn it off above 60 degrees F, the number of hours it can be turned off can be easily determined. Referring to the second sample chart, the periods that are greater than 60 degrees F are highlighted. The number of boxes that are highlighted are counted— in this case there are 132, and then multiplied by 30 (days in a month). For this example, there are 3960 hours the boiler could be turned off.

Similar tables are available for wet bulb temperature.

<u>Limitations</u>: ECM proposals that consider more than one parameter (i.e. dry bulb and wet bulb temperature) are not conveniently used with this form of table. For example, to find the number of annual hours when it is below 55 degrees dry bulb and also below 35 degrees wet bulb is a manual process of comparing the charts cell-by-cell and would be better suited to bin data with coincident wet bulb.

Colorado Springs - Dry Bulb Temperatures

	Jan	Feb	Mar	Apr	May	Jun	Jul	Aug	Sep	Oct	Nov	Dec
0000	25.2	29.2	40.6	50.6	59.6	66.6	69.6	69.6	63.6	53.6	43.2	31.2
0100	23.9	27.9	39.3	49.3	58.3	65.3	68.3	68.3	62.3	52.3	41.9	29.9
0200	22.7	26.7	38.1	48.1	57.1	64.1	67.1	67.1	61.1	51.1	40.7	28.7
0300	21.7	25.7	37.1	47.1	56.1	63.1	66.1	66.1	60.1	50.1	39.7	27.7
0400	20.9	24.9	36.3	46.3	55.3	62.3	65.3	65.3	59.3	49.3	38.9	26.9
0500	20.7	24.7	36.1	46.1	55.1	62.1	65.1	65.1	59.1	49.1	38.7	26.7
0600	21.2	25.2	36.5	46.6	55.6	62.6	65.8	65.6	59.6	49.6	39.2	27.2
0700	22.4	26.4	37.8	47.8	55.8	63.8	66.8	66.8	60.8	50.8	40.4	28.4
0800	24.7	28.7	40.1	50.1	59.1	66.1	69.1	69.1	63.1	53.1	42.7	30.7
0900	27.9	31.9	43.3	53.3	62.3	69.3	72.3	72.3	66.3	56.3	45.9	33.9
1000	31.7	35.7	47.1	57.1	66.1	73.1	76.1	76.1	70.1	60.1	49.7	37.7
1100	35.9	39.9	51.3	61.3	70.3	77.3	80.3	80.3	74.3	64.3	53.9	41.9
1200	39.9	43.9	55.3	65.3	74.3	81.3	84.3	84.3	78.3	68.3	57.9	45.9
1300	42.9	46.9	58.3	68.3	77.3	84.3	87.3	87.3	81.3	71.3	60.9	48.9
1400	44.9	48.9	60.3	70.3	79.3	86.3	89.3	89.3	83.3	73.3	62.9	50.9
1500	45.6	49.6	61.0	71.0	80.0	87.0	90.0	90.0	84.0	74.0	63.6	51.6
1600	44.9	48.9	60.3	70.3	79.3	86.3	89.3	89.3	83.3	73.3	62.9	50.9
1700	43.1	47.1	58.5	68.5	77.5	84.5	87.5	87.5	81.5	71.5	61.1	49.1
1800	40.4	44.4	55.8	65.8	74.8	81.8	84.8	84.8	78.8	68.8	58.4	46.4
1900	37.1	41.1	52.5	62.5	71.5	78.5	81.5	81.5	75.5	65.5	55.1	43.1
2000	33.9	37.9	49.3	59.3	68.3	75.3	78.3	78.3	72.3	62.3	51.9	39.9
2100	31.2	35.2	46.6	56.6	65.6	72.5	75.8	75.6	69.6	59.6	49.2	37.2
2200	28.7	32.7	44.1	54.1	63.1	70.1	73.1	73.1	67.1	57.1	46.7	34.7
2300	26.7	30.7	42.1	52.1	61.1	68.1	71.1	71.1	65.1	55.1	44.7	32.7
2300	31.6	35.6	47.0	57.0	66.0	73.0	76.0	76.0	70.0	60.0	49.6	37.6

Figure 19-1. Sample Weather Data in Month and Hour Format
Source: "Hourly Analysis Program" (HAP), Carrier Corporation, 2003

Example use of table: times when outside air is above 60 degrees.

	Jan	Feb	Mar	Apr	May	Jun	Jul	Aug	Sep	Oct	Nov	Dec
0000	25.2	29.2	40.6	50.6	59.6	66.6	69.6	69.6	63.6	53.6	43.2	31.2
0100	23.9	27.9	39.3	49.3	58.3	65.3	68.3	68.3	62.3	52.3	41.9	29.9
0200	22.7	26.7	38.1	48.1	57.1	64.1	67.1	67.1	60.1	51.1	40.7	28.7
0300	21.7	25.7	37.1	47.1	56.1	63.1	66.1	66.1	60.1	50.1	39.7	27.7
0400	20.9	24.9	36.3	46.3	55.3	62.3	65.3	65.3	59.3	49.3	38.9	26.9
0500	20.7	24.7	36.1	46.1	55.1	62.1	65.1	65.1	59.1	49.1	38.7	26.7
0600	21.2	25.2	36.5	46.6	55.6	62.6	65.8	65.6	59.6	49.6	39.2	27.2
0700	22.4	26.4	37.8	47.8	55.8	63.8	66.8	66.8	60.8	50.8	40.4	28.4
0800	24.7	28.7	40.1	50.1	59.1	66.1	69.1	69.1	63.1	53.1	42.7	30.7
0900	27.9	31.9	43.3	53.3	62.3	69.3	72.3	72.3	66.3	56.3	45.9	33.9
1000	31.7	35.7	47.1	57.1	66.1	73.1	76.1	76.1	70.1	60.1	49.7	37.7
1100	35.9	39.9	51.3	61.3	70.3	77.3	80.3	80.3	74.3	64.3	53.9	41.9
1200	39.9	43.9	55.3	65.3	74.3	81.3	84.3	84.3	78.3	68.3	57.9	45.9
1300	42.9	46.9	58.3	68.3	77.3	84.3	87.3	87.3	81.3	71.3	60.9	48.9
1400	44.9	48.9	60.3	70.3	79.3	86.3	89.3	89.3	83.3	73.3	62.9	50.9
1500	45.6	49.6	61.0	71.0	80.0	87.0	90.0	90.0	84.0	74.0	63.6	51.6
1600	44.9	48.9	60.3	70.3	79.3	86.3	89.3	89.3	83.3	73.3	62.9	50.9
1700	43.1	47.1	58.5	68.5	77.5	84.5	87.5	87.5	81.5	71.5	61.1	49.1
1800	40.4	44.4	55.8	65.8	74.8	81.8	84.8	84.8	78.8	68.8	58.4	46.4
1900	37.1	41.1	52.5	62.5	71.5	78.5	81.5	81.5	75.5	65.5	55.1	43.1
2000	33.9	37.9	49.3	59.3	68.3	75.3	78.3	78.3	72.3	62.3	51.9	39.9
2100	31.2	35.2	46.6	56.6	65.6	72.5	75.8	75.6	69.6	59.6	49.2	37.2
2200	28.7	32.7	44.1	54.1	63.1	70.1	73.1	73.1	67.1	57.1	46.7	34.7
2300	26.7	30.7	42.1	52.1	61.1	68.1	71.1	71.1	65.1	55.1	44.7	32.7

Figure 19.2 Example use of Weather Data in Month and Hour Format
Example use of table: times when outside air is above 60 degrees.

Chapter 20
Pollution and Greenhouse Gases

Since energy use and greenhouse gas emission are directly linked the potential impact of the application of this book is very large.

EMISSION CONVERSION FACTORS BY REGION

It may be desirable to equate electrical energy savings to pollution savings. State and Regional level factors are published and can be used if utility-specific or area-specific emission factors are not available. See **Figure 20-1**.

GREENHOUSE GAS RELATIONSHIP TO ENERGY USE

In the U.S., our greenhouse gas emissions come mostly from energy use. These are driven largely by economic growth, fuel used for electricity generation, and weather patterns affecting heating and cooling needs.
Source: "Greenhouse Gases, Climate Change, and Energy", EIA, 2004

Carbon dioxide emission related to electricity generation:
    81% comes from coal
Source: "Greenhouse Gases, Climate Change, and Energy", EIA, 2004

For total US carbon dioxide emissions, from all sources:
    42% comes from petroleum
    37% comes from coal
    21% comes from natural gas
Source: "Greenhouse Gases, Climate Change, and Energy", EIA, 2004

Region/State	CO_2 lbs/kWh	CH_4 lbs/MWh	N_2O lbs/MWh
New England	**0.98**	**0.0207**	**0.0146**
Connecticut	0.94	0.0174	0.012
Maine	0.85	0.0565	0.027
Massachusetts	1.28	0.0174	0.0159
New Hampshire	0.68	0.0172	0.0141
Rhode Island	1.05	0.0068	0.0047
Vermont	0.03	0.0096	0.0039
Mid Atlantic	**1.04**	**0.0093**	**0.0145**
New Jersey	0.71	0.0077	0.0079
New York	0.86	0.0081	0.0089
Pennsylvania	1.26	0.0107	0.0203
East-North Central	**1.63**	**0.0123**	**0.0257**
Illinois	1.16	0.0082	0.018
Indiana	2.08	0.0143	0.0323
Michigan	1.58	0.0146	0.025
Ohio	1.8	0.013	0.0288
Wisconsin	1.64	0.0138	0.026
West-North Central	**1.73**	**0.0127**	**0.0269**
Iowa	1.88	0.0138	0.0298
Kansas	1.68	0.0112	0.0254
Minnesota	1.52	0.0157	0.0247
Missouri	1.84	0.0126	0.0288
Nebraska	1.4	0.0095	0.0219
North Dakota	2.24	0.0147	0.0339
South Dakota	0.8	0.0053	0.0121
South Atlantic	**1.35**	**0.0127**	**0.0207**
Delaware	1.83	0.0123	0.0227
Florida	1.39	0.015	0.018
Georgia	1.37	0.0129	0.0226
Maryland (*)	1.37	0.0118	0.0206
North Carolina	1.24	0.0105	0.0203
South Carolina	0.83	0.0091	0.0145
Virginia	1.16	0.0137	0.0192
West Virginia	1.98	0.0137	0.0316

Figure 20-1. Pollution-conversion Constants by Region

Source: "Updated State-and Regional-level Greenhouse Gas Emission Factors for Electricity", March 2002, Energy Information Administration

Pollution and Greenhouse Gases

Region/State	CO_2 lbs/kWh	CH_4 lbs/MWh	N_2O lbs/MWh
East-South Central	**1.49**	**0.0128**	**0.024**
Alabama	1.31	0.0137	0.0223
Kentucky	2.01	0.014	0.0321
Mississippi	1.29	0.0132	0.0165
Tennessee	1.3	0.0105	0.0212
West-South Central	**1.43**	**0.0087**	**0.0153**
Arkansas	1.29	0.0125	0.0203
Louisiana	1.18	0.0094	0.0112
Oklahoma	1.72	0.011	0.0223
Texas	1.46	0.0077	0.0146
Mountain	**1.56**	**0.0108**	**0.0236**
Arizona	1.05	0.0068	0.0154
Colorado	1.93	0.0127	0.0289
Idaho	0.03	0.008	0.0033
Montana	1.43	0.0108	0.0227
Nevada	1.52	0.009	0.0195
New Mexico	2.02	0.0131	0.0296
Utah	1.93	0.0134	0.0308
Wyoming	2.15	0.0147	0.0338
Pacific Contiguous	**0.45**	**0.0053**	**0.0037**
California	0.61	0.0067	0.0037
Oregon	0.28	0.0033	0.0034
Washington	0.25	0.0037	0.004
Pacific Non-contiguous	**1.56**	**0.0161**	**0.0149**
Alaska	1.38	0.0068	0.0089
Hawaii	1.66	0.0214	0.0183
U.S. Average	**1.34**	**0.0111**	**0.0192**

*Includes the District of Columbia
CO_2 = Carbon Dioxide
CH_4 = Methane
N_2O = Nitrous Oxide

Figure 20-1. Pollution-conversion Constants by Region (*Continued*)
Source: "Updated State-and Regional-level Greenhouse Gas Emission Factors for Electricity", March 2002, Energy Information Administration

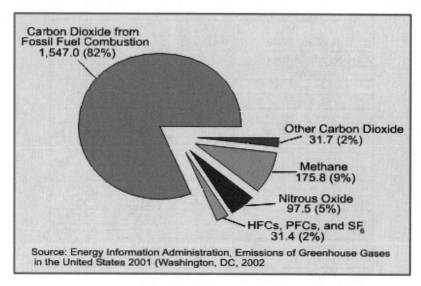

Figure 20-2. Greenhouse Gas Emissions by Gas
2001 (Million Metric Tons of Carbon Equivalent)
Source: "Emissions of Greenhouse Gases in the United States", EIA, 2002

Pollutant	Natural Gas	Oil	Coal
CO_2	117,000	164,000	208,000
CO	40	33	208
NO_x	92	448	457
SO_x	0.6	1,122	2,591
Particulates	7	84	2,744
Mercury	0.000	0.007	0.016

Figure 20-3. Fossil Fuel Emissions
Units: Pounds per Billion Btu of Energy Input
Source: Energy Information Administration - Natural Gas Issues and Trends 1998, Chap 2, Table 2

POLLUTION-CONVERSION TO EQUIVALENT NUMBER OF AUTOMOBILES

Sometimes it is effective to present pollution numbers in terms of automobiles removed from the road.

The numbers vary depending on vintage of the car, miles per gallon, miles per year, and fuel type. The standard for US EPA reporting follows.

Source: US EPA Office of Transportation and Air Quality, April 2000

Notes:
1. These emission factors and fuel consumption rates are averages for the entire in-use fleet. Newer cars and trucks will emit less pollution and use less gasoline; older cars and trucks may emit more pollution and use more gasoline.
2. Carbon dioxide, while not regulated as an emission, is the transportation sector's primary contribution to climate change. Carbon dioxide emissions are directly proportional to fuel economy—**each 1% increase (decrease) in fuel consumption results in a corresponding 1% increase (decrease) in carbon dioxide emissions.**
3. The total annual emissions and fuel consumption are greater for light trucks than was presented in the April 1998 version of this fact sheet. This reflects the increasing trend toward the largest, heaviest light trucks, which currently and in the past have had less stringent emission standards and lower fuel economy than do the lighter light trucks and cars. The new "Tier 2" emission standards taking effect starting with the 2004 model year will bring all light trucks into compliance with the same emission standards as cars (for HC, CO, and NO_x)

Average Vehicle	Total Annual Pollution
Passenger Car	77.1 pounds of hydrocarbons
	575 pounds of carbon monoxide
	38.2 pounds of oxides of nitrogen
	11,450 pounds of CO_2
	581 gallons of gasoline
Light Truck	108 pounds of hydrocarbons
	854 pounds of carbon monoxide
	55.8 pounds of oxides of nitrogen
	16,035 pounds of CO_2
	813 gallons of gasoline

OTHER ENVIRONMENTAL CONSIDERATIONS

Water-cooled Systems

Cost and availability of water is a consideration. Water use is distinctly different from fossil fuel use in that the water we evaporate is re-deposited as rain somewhere else, and our planet as a whole won't run out of water on account of evaporative cooling. However, on a per-location basis, the cost of water may be prohibitive. In one city where electricity cost was relatively low and water cost was relatively high more than 2/3 of the cash value of the energy savings created by water-cooling vs. air-cooling was consumed by the cost of water/sewer.

In some cases, recycled water can be used for evaporative cooling, although it is often found that residual dissolved solids or elevated phosphate contents (or other contents) make water treatment more difficult.

To decrease water use, it is common to treat cooling water with chemicals, to correct pH, to inhibit scale and corrosion to acceptable rates, and to control biological contamination. The chemicals are a mix of acids, bases, phosphates, minerals, and biocides, and can themselves pose water and land pollution risks. Various non-chemical water treatment methods have been attempted with varying success.

Where available, utility credits for evaporated water can lessen the cost of water for evaporation use since the sewer out-flow is usually assumed to equal the building in-flow; when water is used for evaporation very little of it is actually returned to the sewer.

Fluorescent Lighting

A common stabilizing ingredient in fluorescent lights is mercury. Proper disposal is necessary to avoid landfill contamination. Widespread use of fluorescent lighting, especially compact fluorescents, introduces the potential for land and water pollution from improper disposal, which will be a paradigm shift for residential customers who simply 'throw away' light bulbs that have failed.

Chapter 21

Formulas and Conversions

COMMON ENERGY EQUATIONS

COP, EER, kW/Ton
  COP = cooling capacity / work input.
  COP = Btu out (total)/Btu in (external "PAID FOR" only) or work in.
  EER = cooling capacity / work input.
  EER = Btu out (gross cooling)/W-H in or Btuh out/Watts in.

Watts input includes auxiliary equipment such as indoor and outdoor fans. For split systems, or if the indoor fan energy is unknown, ARI 210 adds 1,250 Btu/h per 1,000 cfm and adds by 365 W per 1,000 cfm.

kW/ton = work input/cooling capacity.
COP for chillers includes only the compressor, not the auxiliaries or cooling tower fans.
  EER=COP*3.413=12/kW/ton.
  kW/ton=12/EER=3.517/COP.
  COP=3.517/kW/ton=EER/3.413.
  The constant 3.517 is 12000/3413.

SEER = weighted average of EERs for air-cooled equipment that consider performance improvements in mild weather.

Heat-conversion Factors
  Calorie (Cal)= heat required to raise 1 g water 1 deg C.
  Large Calorie (kCal) = heat required to raise 1 kg water 1 deg C.
  Food Calorie = same units as kCal.

  Btu = heat required to raise 1 lb water 1 degree F.
  Therm = 10^5 Btu.
  Dekatherm = 10^6 Btu.

Affinity Laws—Formulas

$(CFM2/CFM1) = (N2/N1)$

$(GPM2/GPM1) = (N2/N1)$

$(SP2/SP1) = (N2/N1)^2$

$(HP2/HP1) = (N2/N1)^3$

$HP2 = HP1*(SP2/SP1)^{3/2}$

$SP2 = SP1 * (N2/N1)^2$

$HP2 = HP1 (N2/N1)^3$

$HP2 = HP1 (Q2/Q1)^3$

Effect of reducing static pressure:
$HP2 = HP1 \times (SP2/SP1)y^{1.5}$

Electrical Formulas

Motor kW input $= {}^{(HP) * (\%Load) * (0.746)}/_{eff}$

Approx Motor pct load $= {}^{(Measured\ Amps)}/_{(Full\ Load\ nameplate\ Amps)}$

I (amps) $= {}^{kVA * 1000}/_{\sqrt{3} * V}$ ($\sqrt{3}$ is only for 3-phase loads)
I (amps) $= {}^{kW * 1000}/_{PF * \sqrt{3} * V}$ ($\sqrt{3}$ is only for 3-phase loads)

Power Factor:
    $kW = kVA * COS_\Theta$
    $KVAR = KVA * SIN_\Theta$

Right triangle solutions for power factor:
$kVA^2 = kW^2 + kVAR^2$

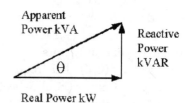

Voltage imbalance (three phase only):
    % imbalance $= \dfrac{(max\ voltage\ on\ any\ line) - (average\ voltage)}{average\ voltage}$

Formulas and Conversions

TO FIND	DC	AC Single Phase	AC 3 PHASE
Amps when Hp is Known	$\dfrac{Hp \times 746}{V \times Eff}$	$\dfrac{Hp \times 746}{V \times Eff \times PF}$	$\dfrac{Hp \times 746}{\sqrt{3} \times V \times Eff \times PF}$
Amps when KW is known	$\dfrac{kW \times 1000}{V}$	$\dfrac{kW \times 1000}{V \times PF}$	$\dfrac{kW \times 1000}{\sqrt{3} \times V \times PF}$
Amps when kVA is known	$\dfrac{kVA \times 1000}{V}$	$\dfrac{kVA \times 1000}{V}$	$\dfrac{kVA \times 1000}{\sqrt{3} \times V}$
KW	$\dfrac{A \times V}{1000}$	$\dfrac{A \times V \times PF}{1000}$	$\dfrac{A \times V \times \sqrt{3} \times PF}{1000}$
KVA	$\dfrac{A \times V}{1000}$	$\dfrac{A \times V}{1000}$	$\dfrac{A \times V \times \sqrt{3}}{1000}$
Hp (output)	$\dfrac{A \times V \times Eff}{746}$	$\dfrac{A \times V \times Eff \times PF}{746}$	$\dfrac{A \times V \times Eff \times \sqrt{3} \times PF}{746}$
Efficiency	$\dfrac{746 \times Output\ Hp}{Input\ Watts}$	$\dfrac{746 \times Hp}{V \times I \times PF}$	$\dfrac{746 \times Hp}{\sqrt{3} \times V \times I \times PF}$
Power Factor		$\dfrac{Input\ Watts}{V \times A}$ Or $\dfrac{kW}{kVA}$	$\dfrac{Input\ Watts}{V \times A \times \sqrt{3}}$ Or $\dfrac{kW}{kVA}$

A = amps Hp = Horsepower
V = Volts Eff = Efficiency

Figure 21-1. Table of Common Electrical Formulas

Load Factor

"Load Factor" is: (average demand / maximum demand).

Commonly applied to electrical demand, but equally applicable to gas or water. Poor load factors are related to high demand charges and improving a low load factor can reduce customer bills by reducing demand charges.

For example, if a customer uses 100,000 kWh in a month and has a 300 kW maximum recorded demand, find the load factor.

Average demand is 300,000/30 days/24 hours = 416 kW
Maximum demand is 900 kW
Load Factor = 416/900 = 46%.

Energy Transport (Circulating Water and Air)

Pump brake horsepower (Bhp) = $\text{gpm} * \text{head} * \text{sp. Gravity} / 3960 * \text{eff}$

Fan Bhp = $\text{cfm} * \text{tsp} / 6356 * \text{eff}$

Differential Temperature

Differential temperature, or Delta-T, or dT, is the motive force for all heat transfer.

Heat transfer will continue to occur until both sides are at equal temperature.

Insulation practices will reduce heat transfer by increasing resistance to heat flow.

Another equally effective method to reduce heat transfer is to reduce the dT.

For a given insulation, half the dT results in half the heat transfer. This is the basis of savings for energy conservation measures like space temperature reset, hot water reset, cool roofs, and passive shading.

Heat Transfer Formulas

The full description of heat transfer includes convection and radiation mechanisms. Only conduction is described in this text, for brevity and since it applies to insulation, a common energy reduction measure.

Conduction heat flow RATE for a building envelope (Btuh):

q = U * A * dT
U = heat transfer coefficient, Btu/SF-degF
A = SF
dT = differential temperature, degrees F

"U" can be for an individual area, or can be a combined, weighted U-value for the "overall" envelope.

For example, if there are 25,000 SF of roof at U = 0.1, 100,000 SF of opaque wall at U = 0.2, 20,000 SF of opaque wall at U = 0.4, and 40,000 SF of glass at U = 0.7, find the overall U-Value

$$\frac{(25{,}000 * 0.1) + (100{,}000 * 0.2) + (20{,}000 * 0.4) + (40{,}000 * 0.7)}{185{,}000}$$

= 0.31 Overall U-Value

Formulas and Conversions

This also demonstrates which of the constituent envelope pieces contributes the most to heat loss/gain.

For this example:

Roof: 2500/58,500 = 4%
Walls: (20,000 + 8000)/58,500 = 48%
Glass: 28,000/58,500 = 48%

Conduction heat flow RATE for glass (Btuh):
$Q = A * SC * SG$
$A = SF$
SC = shading coefficient, a fraction of 1 or less. 1 = no shade
SG = solar gain, Btuh/SF

Conduction heat flow ENERGY (Btu):
$Q = M * C_p * dT$
M = Mass, lb
C_p = Btu/lb-degF
dT = differential temperature, degrees F

Envelope conduction heat transfer ENERGY from Degree Days (Btu):
$Q_{cooling} = U * A * 24 * CDD$

Envelope conduction heat transfer ENERGY from Degree Days (Btu):
$Q_{heating} = U * A * 24 * HDD$

HVAC Formulas and Conversions

Distinction is made between energy (Btu/kWh) and rate (Btuh) since these are commonly confused.

Air Heating RATE (Btuh—sensible):
$q = 1.08 * F_a * cfm * dT$
dT = differential temperature, degrees F
F_a = altitude factor to account for changes in air density
See Item **"Altitude Correction"** in this section for air density ratios (altitude correction factors) at standard temperature, and the **Appendix "Altitude Correction factors at Different Temperatures."**

Total heat flow RATE of air (Btuh):
$4.5 * cfm * dH$

dH = differential Enthalpy, Btu/lb, taken from a psychrometric chart

Water heating RATE (Btuh):
500 * gpm * dT
dT = differential temperature, degrees F

Cooling ENERGY from ton-hours (kWh):
kWh = (Ton-hours) x (kW/ton)
Note: this is an easy calculation, but depends entirely on knowing the number of ton-hours which is not easy to estimate.

Convert air changes per hour (ACH) to cubic feet per minute (CFM)
The volume of air in the enclosure is V (cubic feet)
One air change is one volume-worth of air moved per hour
so
CFM = ACH * V/60
V = volume of the enclosure

Altitude Correction
Air gets thinner at higher altitude and this fact affects many energy consuming equipment items, especially those involving combustion and convective heat transfer. This is significant in two ways:
1. Actual heating and cooling equipment capacities are usually less than nameplate capacities would suggest.
2. Heat transfer surfaces rejecting heat must increase in temperature and surfaces absorbing heat must lower in temperature to achieve the same heat transfer. For heat transfer driven by the refrigeration cycle, where the condensing and evaporating temperatures are created by the compressor, the thermodynamic lift increases and the compressor input power increases, i.e. efficiency is less at higher altitudes.

Fans are affected significantly from altitude. The effect on pumps is negligible—this is because water is largely non-compressible and its density change is small with altitude change. Refrigeration cycle equipment (compressors), and other 'closed' circulating systems are not affected by altitude, other than any heat transfer points.

Formulas and Conversions 333

Some equipment is manufactured with a degree of excess air flow or coil surface area to allow selection without de-rate for the first 2000-3000 feet.

Thinner air (less pounds of it for each cubic foot) means:

- HVAC calculations for air heat transfer are reduced directly as the ratio of <actual air density> to <sea level density>, since the basic formulas are for sea level air density. A table of altitude correction factors for standard temperature follows.
- Fans selected for sea level operation move less pounds of air, and use less brake horsepower.
- HVAC equipment: higher air temperature rise is required for air-cooled condensers, air coils, and dry coolers, with a corresponding higher condensing temperature. A higher air temperature drop is required for air-cooled evaporators, with a corresponding lower evaporator temperature. The approximate relationship for this is 2-4 percent per 1000 feet.
- Less fuel can be added to the mix to maintain the proper air-fuel ratios for combustion. This means combustion equipment is de-rated at altitude, including heaters, boilers, and automobiles. The approximate relationship for this is 4 percent per 1000 feet. Boilers with forced draft fans and engines with superchargers or turbochargers can compensate for this with oversized fans, to artificially raise the atmospheric pressure that the equipment sees.
- Higher air temperature rise is required for air-cooled equipment, with a corresponding higher component or heat exchanger surface temperatures. This affects equipment of all types that is cooled from surrounding air. The approximate relationship for this is 2 percent per 1000 feet for natural convection and 4% per 1000 feet for forced convection. Capacity de-rates apply to variable frequency drives (VFDs) due to the cooling heat sinks and other equipment with air-coils designed for sea-level air density.

Example: A naturally aspirated cast iron boiler rated at 1,000,000 Btuh at sea level, operating at 5,000 feet elevation, would be de-rated by a factor of approximately 5 * 4% = 20%, and have an output of closer to 800,000 Btuh at this altitude.

See also **Appendix "Altitude Correction Factors at Different Temperatures."**

Altitude (ft)	Absolute pressure at 70 degF (psia)	Altitude Correction Factor at 70 degF F_a
0 (Sea Level)	14.7	1.00
1000	14.1	0.96
2000	13.6	0.93
3000	13.1	0.90
4000	12.7	0.86
5000	12.0	0.83
6000	11.8	0.80
7000	11.3	0.77
8000	10.9	0.74
9000	10.5	0.71
10,000	10.1	0.69

Figure 21-2. Air Density Ratios (Altitude Correction Factors)

Humidification Formula
Adding moisture to air:

Humidifier Load (lbs/hr) = 60* Cfm * PCF(air) *delta-M

PCF(air) = lbs/cubic foot air density
delta-M = (lbs moisture/lb dry air)

Note also that each pound of evaporated water used for humidification absorbs about 1000 Btu of heat, either by the heater used to boil it or from the surrounding air.

Properties of Air, Water, Ice
Air density: 0.075 lbs/$_{ft3}$
Specific heat of dry air at STP: 0.24 Btu/$_{lb-degF}$
Density of water at STP: 8.34 lb/gal = 62.4 lbs/$_{ft3}$
Specific heat of water at STP: 1 Btu/$_{lb-degF}$
Specific heat of ice: 144 Btu/$_{lb}$
Volume of water: 7.48gal/$_{ft3}$
7000 grains of moisture = 1 lb of moisture

Specific Heat of Air and Water

Also called heat capacity, this is the heat that can be absorbed by the material for a 1-degree F temperature rise.

Specific heat of water is approximately 1.0 Btu/lbm (pounds mass).
Specific heat of air is approximately 0.24 Btu/lbm at 70 degrees F.

Note: the fact that one pound of water carries four times the heat energy of a pound of air is the reason energy transport using water is more efficient than air: air requires four times the 'pounds per hour' to be circulated, and four times the circulation energy is required to move it.

Latent Heat of Water

This is the phase change heat properties of a material. Water is of interest for boilers and ice makers.

Heat of vaporization for water is approximately 960 Btu/lbm.
Heat of fusion for water is approximately 144 Btu/lbm.

Insulation Formulas

An excellent treatment of practical application of insulation is found in: *Energy Management Handbook*, 6th ed., Chap 15, Turner/Doty.

$R = 1/U$

$U = 1/R$

Thermal Resistivity "R"
The higher the R-value, the higher the insulating value

Units: $(degF * SF * hr)/Btu$

R-values are commonly used to express performance of a chosen thickness. I.e. a 9-inch insulation batt has an R-value of 30 and a 3-inch batt of the same material has an R-value of 11. The "per inch" general parameter has been dropped since the thicknesses have been chosen.

R-values are tabulated for a wide variety of materials in many texts and are not repeated in this text. One source for material R-values of common materials is ASHRAE Fundamentals Handbook.

Overall Heat Transmission Coefficient "U"
The lower the U-value, the higher the insulating value

$$U = 1/R$$

Units: $^{Btu}/_{(hr \ast SF \ast degF)}$

U-Factor is commonly used for evaluating composite layered insulating assemblies. The amount of heat in Btu per hour that is transmitted through one square foot of a surface (wall, floor, roof), per degree F differential temperature.

Evaluating Multiple Layers of Insulation
Calculate the overall effect by adding the individual "R"-values, then the reciprocal ($1/_{Rt}$) is the component U-value. Individual U-values in a layered system cannot be added directly for the composite U_{total}.

Thermal Conductivity "K"
The lower the K-value, the higher the insulating value

$$K = ^{thickness}/_{R}$$

K is the inverse of R. To convert K to R: R= $^{thickness}/K$

Units: $^{(Btu \ast inch)}/_{(hr \ast SF \ast degF)}$

K Factor is heat transfer intensiveness of a material, usually for one inch of thickness.

This is an industry standardized unit used by insulation manufacturers. The "per inch" is a convention that allows different materials to be readily compared.

Note: the K- values are temperature dependent and will vary slightly depending on temperature of service, i.e. the k-value at 100 degrees will be different than at 500 degrees.

OTHER USEFUL FORMULAS

Equivalent Hydraulic Diameter of a Rectangular Shape
D = hydraulic diameter

$$D = 1.3 * \left(\frac{(w*h)^{0.625}}{(w+h)^{0.25}} \right)$$

Duct and Fitting Pressure Losses using "C" Factor
Pressure Loss = $(V/4005)^2 * C$

where:
V = velocity in fpm
C = loss coefficient

Duct and Fitting Pressure Losses using Equivalent Diameters (L/D)
Pressure Loss = $f*(L/D * (V/4005)^2$

where:
f = friction factor
L/D = equivalent diameters

Estimating Unitary HVAC EER from Nameplate Compressor Full Load Amp (FLA) Data

This gives a good approximation for existing equipment, but is not exact. Use manufacturer's literature for Btuh and watts instead of this method, if it is available.

NOTES.
1. Formula is shown for three phase. If single phase, omit the 1.732 factor. FLA = full load amps of the compressor.
2. Power factor PF is approximately 0.8 for motor-compressors.

Step 1. Btuh Cooling Capacity
Nominal capacity inferred from the labeling. The capacity value is usually embedded in the model number. Typical nomenclature:

...024	2 tons
...036	3 tons
...048	4 tons
...060	5 tons
...090	7.5 tons
...120	10 tons
...180	15 tons
...240	20 tons

The capacity is 'nominal' but so is the nameplate FLA

Step 2. Watts
A. Compressor watts: V * FLA * 1.732 * PF
    (Sometimes RLA is shown instead of FLA)
B. Condenser fan watts (for air cooled equipment): add 10% of compressor watts.
C. Evaporator fan watts: 365 watts per 1000 CFM. This is from the ARI rating standard that defines EER ratings.
D. Power Factor (PF) can be assumed to be 0.8 for compressor motors.

Step 3. **EER = Btu/Watts**

Fuel Switching—Electric Resistance Heat vs. Combustion Heat
The economics of this fuel switching option vary by locale.
For heating purposes, the combustion option must include the efficiency loss, so the input is the calculated value. For electric resistance, the efficiency of conversion to heat is assumed to be 100%. Common units can be Btu or therm (for gas heating)
The following example compares natural gas heat to electric heat. Similar approach for other fuels.

Step 1. Convert combustion heat to common units
　　Natural gas cost is $1.00 per therm, or $10.00 per MMBtu
　　Efficiency is 80%
　　Adjusted cost is 10/0.8 = $12.50 per MMBtu

Step 2. Convert electric heat to common units
 Electric cost is $0.07 per kWh
 1 kWh = 3413 Btu
 Efficiency is 100%
 Adjusted cost is $0.07 *1,000,000/3413 = $20.50 per MMBtu

  Factor: ($/kWh) * 293 = $/MMBtu

  Factor: (/kWh) * 29.3 = $/therm

Step 3. Compare fuel cost savings to switch from electric to gas
(20.50 − 12.50)/20.50 = 39% savings to fuel switch for this example.

ENERGY CONVERSION FACTORS

Beware of "M"
In English units, M=1,000
In SI units, M=1,000,000

Other letter designations
D=10
C=100
K=1,000
MM=1,000,000

1 MBtu = 1000 Btu
1 MMBtu = 1,000,000 Btu = 10^6 Btu
1 Quad = 10^{15} Btu

1 MW = 10^6 watts
1 kW = 3413 Btu/h
1 kWh = 3413 Btu

1 ton-hour = 12,000 Btu

1 therm = 100,000 Btu
1 ft^3 natural gas = approx. 1000 Btu at STP = 10^3 Btu at STP
1 ccf natural gas = 100 ft^3 = approx. 100,000 Btu at STP = 10^5 Btu = approx
  1 therm at STP

1 MCF natural gas = 1000 ft³ = approx. 1,000,000 Btu at STP = 10^6 Btu at STP

1 barrel crude oil = 42 gallons = approx. 5,100,000 Btu
1 ton coal = approx. 25,000,000 Btu
1 gallon gasoline = approx. 125,000 Btu
1 gallon #2 fuel oil = approx 140,000 Btu
1 gallon LP gas = approx 95,000 Btu
1 cord of wood = approx 30,000,000 Btu

1 boiler HP = 33,475 Btu/h (33.5 Mbh) **THIS IS OUTPUT**
= 9.809 kW
= 34.5 lbs of water evaporated at 212 degF

1 in. w.c.	0.0360 psi
1 in. w.c.	5.18 psf
1 in. w.c.	0.0733 in. Hg
1 in. w.c.	1.86 mm Hg
1 psi	144 psf
1 psi	2.31 ft. w.c.
1 psi	27.8 in. w.c.
1 psi	2.04 in. Hg
1 psi	51.7 mm Hg
1 atm	14.7 psi
1 atm	408 in. w.c.
1 atm	29.92 in. Hg
1 atm	760 mm Hg

Figure 21-3. Table of Pressure Unit Equivalents

Appendix

GLOSSARY OF TERMS

absorption—A type of cooling system that utilizes heat (often waste heat) to generate the cooling effect. The process is a series of repeating chemical reactions, in contrast to mechanical refrigeration compression cycle cooling.

AEE—Association of Energy Engineers

approach—Heat exchanger context. Since heat exchangers work to bring the temperatures of two fluids closer together, the "approach" is simply how close they can get. The approach value is affected by the amount of surface area of the heat exchanger, the time the two fluids are in contact with each other, and the degree of fouling on the surfaces. As surfaces foul, approach increases. Purchasing extra heat exchange surface area (larger unit) reduces the approach.

ARI—Air-Conditioning and Refrigeration Institute

ASHRAE—American Society of Heating, Refrigerating and Air-Conditioning Engineers

Bhp—Brake hp. Generally used to represent the hp required to drive a machine at the shaft. This value is less than motor horsepower which adds motor efficiency losses

Btu—British thermal unit. A measure of energy in Inch-Pound units. The amount of heat required to raise one pound of water one degree Fahrenheit.

Btuh—Btus per hour

CD—Construction documents. The final phase of completion of design documents.

CFL—Compact fluorescent. A type of lighting.

cfm—Cubic feet per minute. An air flow rate

cfm/SF—Cubic feet per minute air flow per square foot.

COP—Coefficient of performance. Heat out divided by work in, same units. A measure of heating and cooling apparatus efficiency.

CTI—Cooling Tower Institute

Cv—Flow coefficient for a valve, commonly a control valve

CV—Constant volume. An HVAC air system that maintains steady air flow rates at all times, and controls comfort by varying temperatures.

db or dry bulb—Air temperature parameter. The temperature as indicated on an ordinary thermometer, contrasted to "wet bulb" temperature.

DD—Design development. An interim phase of completion for design documents that is after schematic phase and before construction document phase.

DDC—Direct digital control. Refers to microprocessor-based controls that are used to monitor and control equipment and processes, often associated with building systems.

delta T or dT—Differential temperature. Parameter for heat transfer, the difference between inlet and outlet, different sides of a wall, etc.

dP—Differential pressure. A pressure measurement, used in measuring or describing water systems and the energy they consume by pump energy

ECM—Energy conservation measure

EER—A parameter used to evaluate HVAC equipment efficiency Numerically it is the Btu output divided by the watts input. Watts input includes auxiliary equipment such as indoor and outdoor fans.

efficacy—A parameter used to evaluate different types of lighting. Numerically it is the lumen output divided by the watts input

eff—Efficiency. Generally quantified by useful output divided by energy input, expressed as a fraction or percent, less than or equal to 1 or 100%.

EMS—Energy management system. Refers to digital control systems that monitor and control building HVAC systems, lighting systems, and other energy consuming systems.

ft. head or ft. w.c.—A unit of liquid pressure measurements. Refers to the measured vertical rise in an manometer. 1 psi = 2.31 ft w.c.

fpm—Feet per minute. Velocity, usually air.

fps—Feet per second. Velocity, usually water.

gpm—Gallons per minute. A water flow rate.

GSHP—Ground source heat pump. A refrigeration system that provides HVAC heating and cooling, and utilizes the earth mass as a heat sink and heat source.

HID—High intensity discharge. A type of lighting.

hydronic—HVAC system that transports the heating and cooling energy to the points of use by circulating fluid, usually water

HVAC—Heating, ventilating and air conditioning. The general term used to describe systems used to regulate indoor air comfort

hp—Horsepower

IESNA—Illumination Engineering Society of North America

in. w.c.—Inches water column. A low range pressure measurement, used in measuring or describing HVAC air systems and the energy they consume by fan energy. Refers to the vertical rise of a manometer

IRR—Internal rate of return. Economic analysis term

KBtu—Thousands of Btus

kW—Kilowatt. A measure of power or capacity: a thousand watts.

kWh—Kilowatt-hours. A measure of energy

kW/ton—A measure of cooling apparatus efficiency. Watts in divided by cooling tons output.

LCCA—Life cycle cost analysis

LPB—Lighting power budget

Lum—Lumens

MA—Mixed air

MMBtu—Millions of Btus. A measure of energy.

MERV—Minimum efficiency reporting value. A numerical system of rating filters based on a minimum particle size efficiency. A rating of 1 is least efficient; a 16 is the most efficient. See also ASHRAE Standard 52.2.

M&V—Measurement and verification

MW—Megawatt. A measure of power or capacity. A million watts.

NEMA—National Electrical Manufacturers Association

OA—Outside air

O&M or O/M—Operations and maintenance. Manuals.

Pa—Pascals. A unit of pressure.

PC—Performance contract

PF—Power factor. Electrical.

PTAC—Package terminal air conditioner. A through-the-wall unit, as in a hotel.

rH—Relative humidity

RA—Return air

RTU—Rooftop unit. Refers to packaged HVAC equipment located upon the roof of a building

SD—Schematic design. The initial phase of completion for design documents that is before the design development phase

SEER—Seasonally adjusted EER. Based on improved HVAC performance in mild weather

SF—Square feet

SPB—Simple payback. Economic analysis term

TAB—Test and balance

Therm—100,000 Btus. A standard measure of natural gas heating systems, but can apply to any heating system.

Ton (of cooling)—Cooling capacity, e.g. rate of cooling. Convention defines this as the amount of heat required to melt one ton of ice in a 24-hour period. 12,000 Btu/hr cooling rate.

Ton-hour—The factor of tons cooling capacity and hours of duration. A cooling energy unit. 12,000 Btu. Can be equated to kWh if kW/ton is known.

TSP—Total static pressure. Air pressure quantity used to evaluate fan power requirements.

T-8—A type of fluorescent lighting tube. "8-twelfths" of an inch diameter, or 3/4-inch diameter tubes.

T-12—A type of fluorescent lighting tube. "12-twelfths" of an inch diameter, or 1-inch diameter tubes

VAR or VARs—Volt-amp-reactive. Refers to the power use that is non-resistive and does not show up on a watt-meter. This is the symptom of low power factor. The additional current from this "apparent power" (contrasted to "real power) is the source of increased copper losses and voltage drop in distribution systems, i.e. the conductors carry additional amps for no "real" purpose, but suffer the losses just the same.

VAV—Variable air volume. An HVAC air system that maintains a steady air temperature and controls comfort by reducing the system air flow proportionally as cooling demand decreases

VFD—Variable frequency drive. Synonymous to Variable Speed Drive (VSD) and Adjustable Speed Drive (ASD)

W—Watt. A measure of power or capacity. The basic unit of electrical power measure.

W/SF—Watts per square foot

Wb or wet bulb—Wet bulb temperature. A property of air, measured by covering a standard temperature element (the "dry" bulb) with a moist sock and exposing it to the air stream. The amount that wet bulb temperature is less than the 'dry bulb' temperature is a function of the relative humidity of the air.

XA or EA—Exhaust air

CONFLICTING ECMS AND 'WATCH OUTS'

Any engineer's nightmare is to solve one problem and create new ones at the same time. Here are a few of the watch-outs for this field. There will always be more, so tread lightly and seek peer review when you are in new territory. The general answer to all of these is that energy measures are usually system changes and usually more than one thing is affected.

Action	Reaction
Multiple measure savings	Measures affect each other, such that savings are not fully additive. Aggregate savings overstated.
Lighting retrofit savings	Heating energy increases.
Lighting retrofit	Existing heating system was marginally sized and now is inadequate.
Lighting—Occupancy Sensors	Savings assume lights are always left on, but occupants were actually turning their lights off much of the time, so savings are overstated.
Lighting—Daylight Harvesting	Sunlight in abundance will provide good lighting but will also increase heat load in cooling season. Skylights will increase heat loss in heating season.
Refrigeration retrofit for high efficiency	Waste heat was used for heating, and now auxiliary heating elements are needed, increasing energy use.
Chiller savings from reduced head pressure, via lower condenser water temperature	Low efficiency cooling tower requires large increase in tower fan kW, eroding most of the chiller savings. Too much reset can create operational problems for the chiller. Some chillers cannot be reset below 75 degrees. 65-70 deg F is usually safe, and some can accept colder water with excellent reductions in kW.
Chiller savings from increased suction pressure, from increased chilled water temperature set point.	Higher chilled water temperature creates higher apparatus dew point temperature in air handlers, and loss of dehumidification.

(Continued)

Action	Reaction
Condenser water or primary chilled water pump energy savings from variable flow pumping to track chiller load profile.	Changes in average chiller condensing or evaporating temperatures and onset of laminar flow will erode system savings by increasing chiller kW/ton. Still a net gain, but less than calculated by pump energy savings alone.
Evaporative cooling, including cooling towers, evaporative condensers, etc. for compressor savings.	Cost of water and sewer forgotten. This can be a third or half of the energy savings sometimes; It's still worth doing, but remember to subtract those costs from the energy savings.
Condensing boiler retrofit for 90+ efficiency.	High efficiency not achieved because boiler rating depends on low return water temperature, and existing hydronic design is not set up for this.
Power factor correction	Facility with high levels of harmonics can cause premature failure of the capacitors and possible interruption of service.
Air Economizer	Building problems: doors standing open is a common complaint. For buildings with very low break even points, using very cold air for cooling encourages stratification and nuisance freeze stat tripping. Relative humidity swings inside the building for humidity sensitive activities, including electronics, books, and artifacts.
VFD	Savings overestimated using standard cube rule. Prevent this by using square instead of cube and ignoring savings for speeds below 40% speed; assume 40% is the lowest it will go. VFD applied to standard duty motor. Premature failure is likely. Prevent this with Inverter Duty or VFD Grade motors. Damage potential for any splash-lubricated equipment. Prevent this with a minimum speed setting.

(Continued)

Appendix

Action	Reaction
VFD (*Cont'd*)	Damage potential for rotating equipment with a critical frequency. Prevent this with "skip frequency" setting.
	VFD located in hot conditions, near hot equipment or outdoors. Prevent this with remote location or special cooling provisions.
	VFD located upstream of an equipment disconnect. The open circuit scenario will overload the VFD. Prevent this with an interlock kit on the electrical disconnect switch.
Constant Volume HVAC converted to VAV	Low pressure ductwork may not be suitable. Keep pressures low and protect with static pressure switches.
Demand reduction measure savings	Utility 'ratchet' clause can negate savings unless the measure is consistent. A one-time high demand can set the minimum demand charge for up to a year.
Heat recovery savings	Parasitic losses of auxiliary pumps or fans, and system friction added from the equipment in the fluid stream, filters, etc.
Flow savings from increasing delta-T on hydronic systems	The 'signature' of the delta-T is largely fixed by the building loads, e.g. the air handlers and coils, and often the delta-T is limited by the existing building, unless building modifications are also made. Unless coils are replaced to gain more surface area, the system dT can only be increased if both the lower and upper temperatures are extended by the same amount, unless other air system parameters are also changed such as supply air temperature, space temperature, or space relative humidity. For example, changing a 45-55 chilled water coil to 44-56 or changing a 180-160 hot water coil to 185-155, but this impacts the cooling and heating efficiency.
Insulate the floor to reduce winter heat loss	Frozen pipes in the crawl space that were heated from the losses through the floor.
Thermostats turned off in unoccupied areas	Frozen pipes at the perimeter

ENERGY AUDIT TYPES

Walk-Through Audit

Least costly to perform and identifies preliminary energy savings opportunities, but not detailed savings or cost estimates. A visual inspection of the facility is made to identify some of the obvious opportunities, and to determine if more detailed analysis is warranted.

Scoping Audit

Identifies energy conservation measures (ECMs) that appear likely to have a favorable return, usually 5 years or less. Order of magnitude (Note 1) costs and savings are assigned to the suggested measures to allow prioritizing by payback. This audit is normally used to see if further, detailed analysis appears warranted, however the customer can choose this audit as a standalone product if they accept the higher project risk of using the wider tolerance engineering estimates.

Note 1: See **Appendix** item "**Cost Estimating—Accuracy Levels Defined.**"

Investment Grade Audit (IGA)

Provides detailed engineering analysis and is intended to provide sufficient information to support informed choices for capital energy improvements, namely to verify savings, determine costs, and provide economic benefit (cost/benefit ratio) information for each option. The IGA is also appropriate for defining cost/benefit information for manufacturing process or other specialized energy improvement options.

PRESSURE-TEMPERATURE CHARTS FOR REFRIGERANTS

Refrigerants 22, 123, 134A, 404A, 410A, 507

Source: DuPont Fluorochemicals

Figures marked with (") are in. hg.

Saturated Conditions – Pressure (psig)							
Temp. °F	Freon 22 R-22	Suva 123 R-123	Suva 134a R-134a	Suva 404A R-404A	Suva 410A R-410A	Suva 507 R-507	Temp °C
−40	0.6	28.9"	14.8"	4.9	10.8	5.4	−40
−38	1.4	28.8"	13.9"	5.9	12.1	6.4	−39
−36	2.2	28.7"	12.9"	7.0	13.4	7.5	−38
−34	3.1	28.6"	12.0"	8.0	14.8	8.6	−37
−32	4.0	28.5"	10.9"	9.2	16.3	9.8	−36
−30	4.9	28.4"	9.8"	10.3	17.8	11.0	−34
−28	5.9	28.3"	8.7"	11.5	19.4	12.2	−33
−26	6.9	28.2"	7.5"	12.8	21.0	13.5	−32
−24	8.0	28.1"	6.3"	14.1	22.7	14.8	−31
−22	9.1	27.9"	5.0"	15.4	24.5	16.2	−30
−20	10.2	27.8"	3.7"	16.8	26.3	17.6	−29
−18	11.4	27.6"	2.3"	18.3	28.2	19.1	−28
−16	12.6	27.5"	0.8"	19.8	30.2	20.6	−27
−14	13.9	27.3"	0.3	21.3	32.2	22.2	−26
−12	15.2	27.2"	1.1	22.9	34.3	23.8	−24
−10	16.5	27.0"	1.9	24.6	36.5	25.5	−23
−8	17.9	26.8"	2.8	26.3	38.7	27.3	−22
−6	19.4	26.6"	3.6	28.0	41.0	29.1	−21
−4	20.9	26.4"	4.6	29.8	43.4	30.9	−20
−2	22.4	26.2"	5.5	31.7	45.9	32.8	−19
0	24.0	25.9"	6.5	33.7	48.4	34.8	−18
2	25.7	25.7"	7.5	35.7	51.1	36.9	−17
4	27.4	25.4"	8.5	37.7	53.8	39.0	−16
6	29.1	25.2"	9.6	39.8	56.6	41.1	−14
8	31.0	24.9"	10.8	42.0	59.5	43.4	−13

10	32.8	24.6"	11.9	44.3	62.4	45.7	−12
12	34.8	24.3"	13.1	46.6	65.5	48.1	−11
14	36.8	23.9"	14.4	49.0	68.6	50.5	−10
16	38.8	23.6"	15.7	51.5	71.9	53.0	−9
18	40.9	23.2"	17.0	54.0	75.2	55.6	−8
20	43.1	22.9"	18.4	56.6	78.7	58.3	−7
22	45.3	22.5"	19.9	59.3	82.2	61.0	−6
24	47.6	22.1"	21.3	62.0	85.8	63.8	−4
26	50.0	21.7"	22.9	64.8	89.6	66.7	−3
28	52.4	21.2"	24.5	67.7	93.4	69.7	−2
30	55.0	20.8"	26.1	70.7	97.4	72.7	−1
32	57.5	20.3"	27.8	73.8	101.4	75.9	0
34	60.2	19.8"	29.5	76.9	105.6	79.1	1
36	62.9	19.3"	31.3	80.2	109.9	82.4	2
38	65.7	18.7"	33.1	83.5	114.3	85.8	3
40	68.6	18.2"	35.0	86.9	118.8	89.2	4
42	71.5	17.6"	37.0	90.4	123.4	92.8	6
44	74.5	17.0"	39.0	94.0	128.2	96.5	7
46	77.6	16.3"	41.1	97.6	133.0	100.2	8
48	80.8	15.7"	43.2	101.4	138.0	104.1	9
50	84.1	15.0"	45.4	105.3	143.2	108.0	10
52	87.4	14.3"	47.7	109.2	148.4	112.0	11
54	90.8	13.5"	50.0	113.3	153.8	116.2	12
56	94.4	12.8"	52.4	117.4	159.3	120.4	13
58	98.0	12.0"	54.9	121.7	164.9	124.7	14

Saturated Conditions – Pressure (psig)

Temp. °F	Freon 22 R–22	Suva 123 R–123	Suva 134a R–134a	Suva 404A R–404A	Suva 410A R–410A	Suva 507 R–507	Temp °C
60	101.6	11.2"	57.4	126.0	170.7	129.2	16
62	105.4	10.3"	60.0	130.5	176.6	133.7	17
64	109.3	9.4"	62.7	135.0	182.7	138.4	18
66	113.2	8.5"	65.4	139.7	188.9	143.1	19
68	117.3	7.6"	68.2	144.4	195.3	148.0	20
70	121.4	6.6"	71.1	149.3	201.8	153.0	21
72	125.7	5.6"	74.1	154.3	208.4	158.1	22
74	130.0	4.6"	77.1	159.4	215.2	163.3	23
76	134.5	3.5"	80.2	164.6	222.2	168.6	24
78	139.0	2.4"	83.4	169.9	229.3	174.1	26
80	143.6	1.2"	86.7	175.4	236.5	179.6	27
82	148.4	0.0"	90.0	181.0	244.0	185.3	28
84	153.2	0.6	93.5	186.7	251.6	191.1	29
86	158.2	1.2	97.0	192.5	259.3	197.1	30
88	163.2	1.8	100.6	198.4	267.3	203.1	31
90	168.4	2.5	104.3	204.5	275.4	209.3	32
92	173.7	3.2	108.1	210.7	283.6	215.6	33
94	179.1	3.9	112.0	217.0	292.1	222.1	34
96	184.6	4.6	115.9	223.4	300.7	228.7	36
98	190.2	5.3	120.0	230.0	309.5	235.4	37
100	195.9	6.1	124.2	236.8	318.5	242.3	38
102	201.8	6.9	128.4	243.6	327.7	249.3	39
104	207.7	7.7	132.7	250.6	337.1	256.5	40
106	213.8	8.5	137.2	257.8	346.7	263.8	41
108	220.0	9.4	141.7	265.1	356.5	271.2	42

110	226.4	10.3	146.4	272.5	366.4	278.8	43
112	232.8	11.2	151.1	280.1	376.6	286.6	44
114	239.4	12.1	156.0	287.9	387.0	294.5	46
116	246.1	13.1	160.9	295.8	397.6	302.6	47
118	253.0	14.1	166.0	303.8	408.4	310.8	48
120	260.0	15.1	171.2	312.1	419.4	319.2	49
122	267.1	16.1	176.5	320.4	430.7	327.8	50
124	274.3	17.2	181.8	329.0	442.1	336.5	51
126	281.7	18.3	187.4	337.7	453.8	345.4	52
128	289.2	19.4	193.0	346.6	465.8	354.5	53
130	296.9	20.6	198.7	355.6	477.9	363.8	54
132	304.7	21.7	204.6	364.9	490.3	373.2	56
134	312.6	22.9	210.6	374.3	503.0	382.9	57
136	320.7	24.2	216.7	383.9	515.9	392.7	58
138	329.0	25.5	222.9	393.7	529.1	402.7	59
140	337.4	26.8	229.2	403.7	542.5	413.0	60
142	345.9	28.1	235.7	413.9	556.2	423.4	61
144	354.6	29.5	242.3	424.3	570.2	434.1	62
146	363.5	30.9	249.0	434.9	584.5	445.0	63
148	372.5	32.3	255.9	445.7	599.0	456.1	64
150	381.7	33.8	262.9	456.8	613.88	467.4	66

Appendix

Refrigerant 717—Ammonia

Source: Extol of Ohio, Inc.

Values in bold italic are in. hg.

TEMP. °F	PSIG	TEMP. °F	PSIG	TEMP. °F	PSIG	TEMP. °F	PSIG
-60	*18.6*	-14	6.7	31	46.3	76	128.3
-58	*17.8*	-13	7.3	32	47.6	77	130.7
-57	*17.4*	-12	7.9	33	48.9	78	133.2
-56	*17*	-11	8.4	34	50.2	79	135.8
-55	*16.6*	-10	9	35	51.6	80	138.3
-54	*16.2*	-9	9.6	36	52.9	81	140.9
-53	*15.7*	-8	10.3	37	54.3	82	143.6
-52	*15.3*	-7	10.9	38	55.7	83	146.3
-51	*14.8*	-6	11.6	39	57.2	84	149
-50	*14.3*	-5	12.2	40	58.6	85	151.7
-49	*13.8*	-4	12.9	41	60.1	86	154.5
-48	*13.3*	-3	13.6	42	61.6	87	157.3
-47	*12.8*	-2	14.3	43	63.1	88	160.1
-46	*12.2*	-1	15	44	64.7	89	163
-45	*11.7*	0	15.7	45	66.3	90	165.9
-44	*11.1*	1	16.5	46	67.8	91	168.9
-43	*10.6*	2	17.2	47	69.5	92	171.9
-42	*10*	3	18	48	71.1	93	174.9
-41	*9.3*	4	18.8	49	72.8	94	178
-40	*8.7*	5	19.6	50	74.5	95	181.1
-39	*8.1*	6	20.4	51	76.2	96	184.2
-38	*7.4*	7	21.2	52	78	97	187.4
-37	*6.8*	8	22.1	53	79.7	98	190.6
-36	*6.1*	9	22.9	54	81.5	99	193.9
-35	*5.4*	10	23.8	55	83.4	100	197.2
-34	*4.7*	11	24.7	56	85.2	101	200.5
-33	*3.9*	12	25.6	57	87.1	102	203.9
-32	*3.2*	13	26.5	58	89	103	207.3
-31	*2.4*	14	27.5	59	90.9	104	210.7
-30	*1.6*	15	28.4	60	92.9	105	214.2
-29	*0.8*	16	29.4	61	94.9	106	217.8
-28	*0*	17	30.4	62	96.9	107	221.3
-27	0.4	18	31.4	63	98.9	108	225
-26	0.8	19	32.5	64	101	109	228.6
-25	1.3	20	33.5	65	103.1	110	232.3
-24	1.7	21	34.6	66	105.3	111	236.1
-23	2.2	22	35.7	67	107.4	112	239.8
-22	2.6	23	36.8	68	109.6	113	243.7
-21	3.1	24	37.9	69	111.8	114	247.5
-20	3.6	25	39	70	114.1	115	251.5
-19	4.1	26	40.2	71	116.4	116	255.4
-18	4.6	27	41.4	72	118.7	117	259.4
-17	5.1	28	42.6	73	121	118	263.5
-16	5.6	29	43.8	74	123.4	119	267.6
-15	6.2	30	45	75	125.8	120	271.7

COST ESTIMATING—ACCURACY LEVELS DEFINED

Source: AACE International Recommended Practice No. 18R-97 "Cost Estimating Classification System—as Applied in Engineering, Procurement, and Construction for the Process Industries," 1997. Reprinted with the permission of AACE International, 209 Prairie Ave., Suite 100, Morgantown, WV 25601 USA. http://www.aacei.org, copyright 2007 © by AACE International; all rights reserved.

AACE Classification Standard	ANSI Standard Z94.0	AACE Pre-1972	Association of Cost Engineers (UK) ACostE	Norwegian Project Management Association (NFP)	American Society of Professional Estimators (ASPE)
Class 5	Order of Magnitude Estimate -30/+50	Order of Magnitude Estimate	Order of Magnitude Estimate Class IV -30/+30	Concession Estimate	Level 1
				Exploration Estimate	
				Feasibility Estimate	
Class 4	Budget Estimate -15/+30	Study Estimate	Study Estimate Class III -20/+20	Authorization Estimate	Level 2
Class 3		Preliminary Estimate	Budget Estimate Class II -10/+10	Master Control Estimate	Level 3
Class 2	Definitive Estimate -5/+15	Definitive Estimate	Definitive Estimate Class I -5/+5	Current Control Estimate	Level 4
Class 1		Detailed Estimate			Level 5
					Level 6

(INCREASING PROJECT DEFINITION)

Source: AACE International Recommended Practice No. 17R-97 "Cost Estimate Classification System," 1997. Reprinted with the permission of AACE International, 209 Prairie Ave., Suite 100, Morgantown, WV 25601 USA. http://www.aacei.org, copyright 2007 © by AACE International; all rights reserved.

ESTIMATE CLASS	LEVEL OF PROJECT DEFINITION Expressed as % of complete definition	END USAGE Typical purpose of estimate	EXPECTED ACCURACY RANGE Typical +/- range relative to best index of 1 [a]
Class 5	0% to 2%	Screening or Feasibility	4 to 20
Class 4	1% to 15%	Concept Study or Feasibility	3 to 12
Class 3	10% to 40%	Budget, Authorization, or Control	2 to 6
Class 2	30% to 70%	Control or Bid/ Tender	1 to 3
Class 1	50% to 100%	Check Estimate or Bid/ Tender	1

[a] If the range index value of "1" represents +10/-5%, then an index value of 10 represents +100/-50%.

Appendix

SIMPLE PAYBACK VS. INTERNAL RATE OF RETURN (IRR)

Alternate Method to Calculate Internal Rate of Return (IRR)

Internal rate of return (IRR) is that interest rate that produces a net present worth of zero for a series of cash flows. It can also be defined as *the interest rate where the present worth of the project cost equals the present worth of the savings,* or *the interest rate where P/A=1.*

The IRR is compared to the cost of borrowed money (the hurdle rate), and measures with an IRR are normally not pursued unless the IRR is greater than the hurdle rate. Simple Payback Period is a more common representation of project merit, although IRR is more accurate since it includes the time value of money. Simple payback can be equated to IRR if the payback period and the measure life are known.

Example IRR Calculation
Investment $200,000
Life of measure 10 years
Annual savings $40,000
5-year simple payback period (SPP)

First Cost = present worth of the savings at some interest value
P = A (P/A, i?, 10 yrs)
200,000 = 40,000 (P/A, i?, 10 yrs)
factor P/A (simple payback)
200,000/40,000 = (P/A, i?, 10 yrs)
5 = (P/A, i?, 10 yrs)
now, pick through the interest tables for (P/A, i?, 10 yrs) = 5.0, and that is the equivalent interest rate or IRR (internal rate of return). In this case, IRR = 15%

Simple payback period (SPP) is the cost/benefit ratio of a project. SPP and IRR can be equated, provided the life of the measure is known.

Derivation: Internal Rate of Return (IRR) is that interest rate where the present worth of the savings is equal to the initial investment

$P = A * (P/A, i, n)$

$$P = A * \frac{(1+i)^n - 1}{i(1+i)^n}$$

so, for some value if **i**,

$$P/A \text{ (simple payback)} = \frac{(1+i)^n - 1}{i(1+i)^n}$$

Equivalent IRR for Various Payback Periods

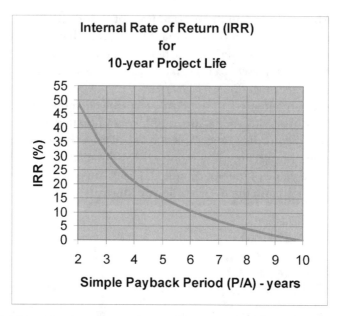

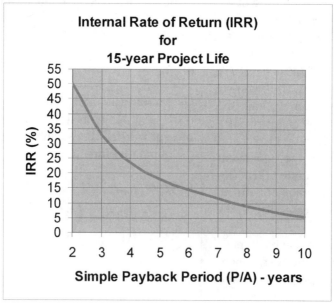

Appendix

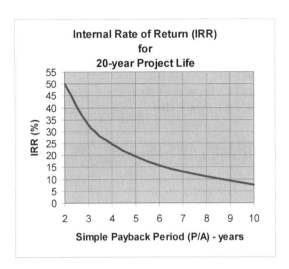

The true benefit of this relationship is shown below. While SPP is the common metric, it is the true time value of money that is most important for viable economic decisions. Using this table, the desired IRR will determine the maximum SPP for a given project life.

SIMPLE PAYBACK	ECM PROJECT LIFE			
	5 YEAR	10 YEAR	15 YEAR	20 YEAR
	IRR	IRR	IRR	IRR
2	40.0%	49.0%	50.0%	50.0%
3	**20.0%**	31.0%	33.0%	33.0%
4	8.0%	**21.0%**	23.5%	24.6%
5		15.0%	**18.0%**	19.4%
6		10.6%	14.5%	**15.8%**
7		7.0%	11.5%	13.0%
8		4.3%	9.0%	11.0%
9		1.8%	7.1%	9.2%
10			5.5%	7.7%

Simple Payback vs. Internal Rate of Return
Maximum payback periods noted
that achieve greater than 15% rate of return
for various project lifespans

HEAT LOSS FROM UNINSULATED HOT PIPING AND SURFACES

Sources: 180 deg F and higher: ASHRAE Fundamentals 1993 © American Society of Heating, Refrigerating and Air-Conditioning Engineers, Inc., www.ashrae.org.

140 deg F bare piping loss data: "Engineered Plumbing Design," American Society of Plumbing Engineers, 1982.

160 deg F bare piping loss interpolated

120 deg F bare piping loss extrapolated

	120 degF	140	160	180	280	380	480	580	1080	
Pipe (in)										
1	35	53	71	89	221	396	623	911	3742	Btuh per foot
2	61	91	121	152	378	681	1076	1581	6607	
3	86	129	172	216	539	973	1542	2271	9562	
4	109	163	217	272	689	1228	1949	2876	12,179	
6	156	233	310	387	969	1758	2797	4138	17,667	
8	229	299	396	493	1236	2246	3580	5305	22,758	
Flat, vertical				212	533	973	1559	2321	10,231	Btuh per SF
Flat, horizontal, facing up				235	586	1061	1683	2485	10,606	
Flat, horizontal, facing down				184	465	861	1400	2113	9754	

HEAT LOSS FROM INSULATED PIPING

Source: Engineering Toolbox Calculator

Assumes 0.1 mph wind for indoor piping, and insulation K=0.25
As a general rule of thumb, the heat loss for insulated vs. un-insulated pipe is a 10:1 reduction in heat loss. Units are Btuh per foot.

Pipe (in)	Insulation (in)	120	140	160	180	200 deg F
1	1	5	8	10	12	14
2	1	9	12	15	19	22
3	1	11	16	21	25	30
4	1	14	20	26	32	37
4	2	9	13	16	20	23
6	1	20	28	36	44	52
6	2	12	17	22	27	31
8	1	26	36	46	57	67
8	2	15	21	27	33	39

Appendix

DUCT FITTING LOSS COEFFICIENTS

This table shows a few common fittings with high transport energy losses. For a complete listing of duct fittings, refer to the sources at the end of the chart, and to *ASHRAE Fundamentals Handbook*.

Type	Description	Equiv. duct diam. (L/D)	Velocity Heads (C)
REDUCERS	Blunt 90 deg reduction (25% reduction)	*12* *	0.25
	Blunt 90 deg reduction (75% reduction)	*25* *	0.5
	Reduction 45 degrees per side	*10* *	0.2
	Reduction 30 degrees per side	*5* *	0.1
	Reduction 15 degrees per side	*2.5* *	0.05
	Blunt 90 deg expansion (V1/V2=0.4)	*25* *	0.5
	Blunt 90 deg expansion (V1/V2=0.8)	*35* *	0.7
	Expansion 10 degrees per side	*30* *	0.6
	Expansion 15 degrees per side	*50* *	1.0
	Expansion 45 degrees per side	*75* *	1.5
DAMPERS	**Fire Dampers**		
	Type "A" Fire Damper (retracted curtain pocket as obstruction in the airstreams)	*75* *	1.5
	Type "B" Fire Damper (retracted curtain pocket out of the airstreams)	*10* *	0.2
	Control Dampers or Balancing Dampers (full open)		
	flat, stamped blade shape	*40* *	0.8
	air foil blade shape	*15* *	0.3
ENTRANCES and EXITS	Entrance - abrupt	*18* *	0.35
	Entrance - bell mouth	*1.5* *	0.03
	Exit - all	all pressure lost	all pressure lost
FAN DISCHARGE	Fan blast outlet area 60% of fan connection size, forward curved fan		
	Discharge Transition		
	12% of ideal effective duct (Note 5)		0.7
	25% of ideal effective duct		0.3
	50% of ideal effective duct		0.1
	100% of ideal effective duct		0.0
	Elbow Too Close to Discharge		
	12% of ideal effective duct (Note 5)		1.6
	25% of ideal effective duct		1.2
	50% of ideal effective duct		0.5
	100% of ideal effective duct		0.0
FAN INLET	**No Vanes**		
	Elbow @ fan inlet, no straight section		2.0
	Elbow @ fan inlet, 2D straight section		1.2
	Elbow @ fan inlet, 5D straight section		0.5
	With Vanes		
	Elbow @ fan inlet, no straight section		0.8
	Elbow @ fan inlet, 2D straight section		0.5
	Elbow @ fan inlet, 5D straight section		0.2
	0% obstructed fan inlet		0.0
	25% obstructed fan inlet		0.8
	50% obstructed fan inlet		1.6
	Fan inlet too close to cabinet wall, 0.75x fan diameter to the wall		0.2
	Fan inlet too close to cabinet wall, 0.5x fan diameter to the wall		0.4

Type	Description	Equiv. duct diam. (L/D)	Velocity Heads (C)
ELBOWS	**Radius Rectangular Elbow** (R/D=1, no vanes, W/D 2.0 - 3.0, average bend)		
	90 deg ell	15	*0.3* *
	45 deg ell	7.5	*0.15* *
	Rectangular Square Elbow (equal dimensions in/out, with vanes. Increase 4x if no vanes)		
	90 degree ell, vanes	15	0.7
	90 degree ell, no vanes	60	1.2
	90 degree offset "Z" - 2 ells, vanes	45	*0.9* *
	90 degree offset - 2 ells, no vanes	*130* *	2.6
	90 degree offset - 2 ells, different planes	*170* *	3.4
	45 degree ell, no vanes	*15* *	0.3
	Round Elbows (R/D=1, no vanes)		
	90 degree ell, smooth	10	0.2
	90 degree ell, 3-piece or 5-piece	25	0.4
	90 offset - 2 ells	75	*1.5* *
	45 degree ell, smooth	5	0.13
	45 degree ell, 3-piece	15	0.24
	90 degree offset - 2 ells	75	*1.5* *
	45 degree offset, 3-piece	45	*0.9* *
	Round mitered ell (no vanes)		
	90 degree ell	65	1.2
	90 degree offset	195	*3.9* *
	45 degree ell	35	0.34
	45 degree offset	105	*2.1* *

Adapted from Carrier System Design Manual,1974 and SMACNA HVAC Systems Duct Design Manual,1990

Notes:
1. Values marked with (*) were calculated with the approximate relationship L/D=50C

2. L/D = equivalent diameters, equal units, where

   D = hydraulic diameter = $D = 1.3 * \left(\frac{(w*h)^{0.625}}{(w+h)^{0.25}} \right)$

3. Pressure Loss Formula: Duct fitting and system pressure losses, using "C" Factor

   Pressure Loss = $(V/4005)^2 * C$
   where:
   V = velocity in fpm
   C = loss coefficient

   Duct and fitting pressure losses using equivalent diameters (L/D)
   Pressure Loss = $f(L/D) * (V/4005)^2$
   where:
   f = friction factor
   L/D = equivalent diameters

4. For fittings with a change in velocity, use the difference of the inlet and outlet velocity head.
5. Fan discharge velocity profile is non-uniform and the fan system will incur loss unless duct geometry allows sufficient length of straight duct to establish a uniform velocity profile before take-offs or bends.
To calculate "100% effective duct length", assume a minimum of 1-1/2 duct diameters for each 2500 fpm or less, and add 1 duct diameter for each 1000 fpm over that.

EVAPORATION LOSS FROM WATER IN HEATED TANKS

Btu/hr-SF

Source, "Energy Tips," Office of Industrial Technologies (ITP), US DOE Office of Energy Efficiency and Renewable Energy.

Liquid Temp, deg F	65 deg F ambient air temp	75 deg F ambient air temp	85 deg F ambient air temp	95 deg F ambient air temp	105 deg F ambient air temp
110	244	222	200	177	152
130	479	452	425	397	369
150	889	856	822	788	754
170	1608	1566	1524	1482	1440
190	2900	2845	2790	2737	2684

Note 1: These losses assume still air over the tanks. For push air tanks, or outdoor tanks, adjust for wind speed as follows:
Source: ASHRAE Applications Handbook, 1995 © American Society of Heating, Refrigerating and Air-Conditioning Engineers, Inc., www.ashrae.org.

Air Speed	Factor for still-air evaporation loss
5 mph (440 fpm)	X 1.25
10 mph (880 fpm)	X 2.0

Note 2: When tank evaporative loss requires make-up water, this requires even more heat for the make-up. With 60 deg make-up, the added heat for make-up is approximately:

Tank Temp, deg F	Percent Added Heat Loss (60 deg F make-up)
100	4%
140	7%
180	11%

BIN WEATHER DATA FOR 5 CITIES (DRY BULB)

Source: Bin Maker Pro 2.0.1, Interenergy Software, Gas Technology Institute.

Sample Bin Weather Data. Dry Bulb Bins
Source: Bin Maker Pro 2.0.1, Interenergy Software/ Gas Technology Institute
work week hours shown are 6a-8p M-F, and 10a-5p Sa,Su

Mid-pts	DB (F)	Climate Zone 1 Anchorage AK		Climate Zone 2 Colorado Springs, CO		Climate Zone 3 Portland OR		Climate Zone 4 Atlanta GA		Climate Zone 5 Tampa FL	
		24x7 Total Hrs	work week Total Hrs	24x7 Total Hrs	work week Total Hrs	24x7 Total Hrs	work week Total Hrs	24x7 Total Hrs	work week Total Hrs	24x7 Total Hrs	work week Total Hrs
117.5	115 to 120										
112.5	110 to 115										
107.5	105 to 110										
102.5	100 to 105										
97.5	95 to 100							9	9	5	5
92.5	90 to 95			37	36	25	25	56	52	238	238
87.5	85 to 90			100	99	54	51	196	187	644	621
82.5	80 to 85			285	270	146	131	758	681	1389	1124
77.5	75 to 80	18	16	369	335	246	216	768	515	1701	796
72.5	70 to 75	72	65	461	360	462	391	1314	571	1526	671
67.5	65 to 70	100	86	539	353	496	361	885	403	908	367
62.5	60 to 65	474	366	865	460	1049	569	1027	567	937	366
57.5	55 to 60	937	538	813	384	1288	611	790	415	504	209
52.5	50 to 55	1069	552	744	372	1259	696	673	321	454	197
47.5	45 to 50	721	351	729	380	1436	679	641	285	220	66
42.5	40 to 45	543	277	657	325	1115	522	436	210	106	36
37.5	35 to 40	919	414	869	398	796	331	560	259	101	33
32.5	30 to 35	740	401	693	295	253	97	323	119	27	11
27.5	25 to 30	867	467	561	245	61	29	181	77		
22.5	20 to 25	504	280	399	153	60	27	72	40		
17.5	15 to 20	778	420	302	101	14	4	64	25		
12.5	10 to 15	418	202	134	83			7	4		
7.5	5 to 10	250	129	95	51						
2.5	0 to 5	194	117	81	30						
-2.5	-5 to 0	66	30	24	8						
-7.5	-10 to -5	68	19	3	2						
-12.5	-15 to -10	22	10								
-17.5	-20 to -15										

Appendix

ALTITUDE CORRECTION FACTORS AT DIFFERENT TEMPERATURES

Air Density Correction Factors

Units psia
Corrected Pressure = 14.7/factor
Source of data: Greenheck Corporation, www.greenheck.com

K=1000 ft elev.

Air Temp degF	0	1K	2K	3K	4K	5K	6K	7K	8K	9K	10K
0	0.87	0.90	0.94	0.97	1.01	1.05	1.08	1.13	1.17	1.22	1.26
50	0.96	1.00	1.04	1.08	1.11	1.15	1.20	1.24	1.30	1.34	1.40
70	1.00	1.04	1.08	1.12	1.16	1.22	1.25	1.30	1.35	1.40	1.45
100	1.06	1.10	1.14	1.18	1.22	1.27	1.32	1.37	1.42	1.48	1.54
200	1.25	1.29	1.34	1.40	1.44	1.50	1.56	1.61	1.68	1.75	1.81
300	1.43	1.49	1.54	1.60	1.66	1.72	1.79	1.86	1.93	2.01	2.08
400	1.62	1.68	1.75	1.81	1.88	1.94	2.03	2.09	2.19	2.27	2.37
500	1.81	1.88	1.95	2.02	2.10	2.18	2.26	2.35	2.44	2.54	2.63
600	2.00	2.07	2.15	2.23	2.31	2.40	2.50	2.59	2.69	2.84	2.91
700	2.19	2.27	2.35	2.44	2.53	2.63	2.73	2.83	2.94	3.07	3.17
800	2.38	2.48	2.57	2.67	2.76	2.86	2.98	3.09	3.21	3.33	3.45
900	2.56	2.68	2.78	2.87	2.97	3.07	3.20	3.33	3.46	3.58	3.71
1000	2.76	2.87	2.99	3.09	3.20	3.31	3.45	3.59	3.73	3.85	4.00

ENERGY USE INTENSITY (EUI)—PER SF—BY FUNCTION AND SIZE

Sources: Except as noted: EIA / CBECS (Commercial Building Energy Consumption) 2003, Raw Data Files Sorted for Functions Shown.
Source 2. Local sub meter data

"Other" and "Vacant" entries omitted.
Units are kBtu/SF-yr

Category	Function	kBtu/SF-yr
Assembly	Entertainment/culture, recreation, social/meeting, library, other public assembly	94
	Library	138
Computer Data Center	All energy use within the data center (source 2)	1170
	Computers and cooling units only (source 2)	865
Education	College/university buildings	155
	K-12 school	72
Food Sales	Grocery store/food market	214
Food Service	Fast food	451
	Full service restaurant/cafeteria	231
	Catering, bars, donut/bagel shops, coffee shops, ice cream shops	193
Healthcare	Hospital	249
	Outpatient	95
	Assisted living/nursing home	125
Lodging	0 - 50,000 SF	83
	> 50,000 SF	100
Office	0 - 50,000 SF	82
	> 50,000 SF	102
Retail	0-50,000 SF > 50,000 SF	71 82
Warehouse	Distribution/shipping center, non-refrigerated warehouse, self-storage	45
	Refrigerated storage	98
Worship	Church	44

ENERGY USE INTENSITY (EUI)—PER SF—
BY FUNCTION AND CLIMATE ZONE

Source: EIA/CBECS (Commercial Building Energy Consumption) 2003, Table C10, National Averages. "Other" and "Vacant" entries omitted. Units are kBtu/SF-yr

PRINCIPAL BUILDING ACTIVITY	1	2	3	4	5
Education	91.6	85.2	93.5	76.6	72.6
Food Sales	187.5	196.7	Q	189.1	Q
Food Service	230.1	238.7	247.6	305.4	247.1
Health Care	202.4	205.8	191.4	171.9	167.7
Inpatient	246.1	283.3	277.2	215.0	236.8
Outpatient	148.5	101.5	76.0	70.9	79.6
Lodging	90.1	132.1	97.1	81.4	87.6
Retail (Other Than Mall)	103.0	73.5	77.7	50.7	92.0
Office	90.7	114.9	95.4	69.3	83.2
Public Assembly	102.2	90.8	87.3	111.1	65.8
Public Order and Safety	Q	135.8	Q	Q	Q
Religious Worship	62.9	46.9	52.8	29.4	38.2
Service	100.4	71.2	99.5	50.3	64.6
Warehouse and Storage	45.7	76.1	49.5	30.1	16.9

ENERGY USE INTENSITY (EUI)—PER SF—
BY FUNCTION, CLIMATE ZONE, AND SIZE

Sorted by primary business activity, then by climate zone, then by size. Source: U.S. EIA CBECS Survey Results, 2003 Raw Data Set.

Notes: For some categories, "Small" and "Large" distinction made, where small is 0-50KSF and large>50KSF. CBECS staff recommends caution with calculations with less than 20 records, and so these are highlighted.

The numbers in these or any CEBCS tables are not absolutes. Plus/Minus tolerances for the values that follow have not been calculated, since they are from combined CEBCS tables. As a general rule of thumb from the Author, applying a +/- 20% tolerance to figures that are comprised of 20 or more records should provide a reasonable benchmark.

Manual Alterations to Published Raw Data Set
1. Removed "vacant" and "other" categories since these are not applicable for Energy Auditing.
2. Removed all records reporting zero energy use as either reporting errors or not applicable for Energy Auditing.

366   Commercial Energy Auditing Reference Handbook

	Office kBtu/SF-yr	# Records	Lab kBtu/SF-yr	# Records	Warehouse kBtu/SF-yr	# Records	Food Sales kBtu/SF-yr	# Records
Small, climate zone 1	89	117						
Large, climate zone 1	99	42						
Small, climate zone 2	100	150						
Large, climate zone 2	133	101						
Small, climate zone 3	80	90						
Large, climate zone 3	119	97						
Small, climate zone 4	66	137						
Large, climate zone 4	74	70						
Small, climate zone 5	69	87						
Large, climate zone 5	83	54						
All sizes, climate zone 1	92	159	221	7	55	71	188	30
All sizes, climate zone 2	116	251	390	17	81	93	197	30
All sizes, climate zone 3	102	187	472	7	47	69	251	15
All sizes, climate zone 4	70	207	227	10	23	98	189	33
All sizes, climate zone 5	75	141	206	2	17	88	189	16
All climate zones, All sizes	93	976	305	43	45	423	200	124
All climate zones, Small	82	581						
All climate zones, Large	102	395						

Appendix

	kBtu/SF-yr Public Order and Safety	# Records	kBtu/SF-yr Outpatient Health Care	# Records	kBtu/SF-yr Refrigerated Ware-house	# Records	kBtu/SF-yr Religious Worship	# Records
Small, climate zone 1								
Large, climate zone 1								
Small, climate zone 2								
Large, climate zone 2								
Small, climate zone 3								
Large, climate zone 3								
Small, climate zone 4								
Large, climate zone 4								
Small, climate zone 5								
Large, climate zone 5								
All sizes, climate zone 1	114	19	149	25			63	41
All sizes, climate zone 2	136	27	101	40	64	3	47	104
All sizes, climate zone 3	93	15	76	27	79	6	53	37
All sizes, climate zone 4	117	16	71	29	134	4	29	88
All sizes, climate zone 5	91	8	80	23	119	5	38	41
					103	1		
All climate zones, All sizes	116	85	95	144	98	20	43	311
All climate zones, Small								
All climate zones, Large								

	Public Assembly kBtu/SF-yr	# Records	Education kBtu/SF-yr	# Records	Food Service kBtu/SF-yr	# Records	Inpatient Health Care kBtu/SF-yr	# Records
Small, climate zone 1	69	43	99	43				
Large, climate zone 1	151	18	89	59				
Small, climate zone 2	78	49	88	83				
Large, climate zone 2	126	16	81	91				
Small, climate zone 3	76	19	72	29				
Large, climate zone 3	89	16	99	60				
Small, climate zone 4	78	55	75	118				
Large, climate zone 4	138	15	78	56				
Small, climate zone 5	44	33	60	73				
Large, climate zone 5	98	10	84	35				
All sizes, climate zone 1	102	61	92	102	230	42	259	34
All sizes, climate zone 2	91	65	83	174	239	59	273	44
All sizes, climate zone 3	86	35	94	89	248	32	244	31
All sizes, climate zone 4	102	70	77	174	305	66	218	44
All sizes, climate zone 5	56	43	76	109	247	43	248	37
All climate zones, All sizes	94	278	83	648	258	242	249	217
All climate zones, Small	70	199	78	346				
All climate zones, Large	125	79	86	302				

Appendix

	kBtu/SF-yr		kBtu/SF-yr		kBtu/SF-yr		kBtu/SF-yr	
	Nursing	# Records	Lodging	# Records	Retail Other Than Mall	# Records	Service	# Records
Small, climate zone 1			90	27	108	50		
Large, climate zone 1			69	20	83	11		
Small, climate zone 2			97	38	71	67		
Large, climate zone 2			149	23	85	13		
Small, climate zone 3			65	25	73	53		
Large, climate zone 3			93	22	91	12		
Small, climate zone 4			73	43	48	87		
Large, climate zone 4			84	24	61	15		
Small, climate zone 5			84	25	85	32		
Large, climate zone 5			96	13	103	14		
All sizes, climate zone 1	146	12	77	47	103	61	104	86
All sizes, climate zone 2	137	30	130	61	73	80	72	107
All sizes, climate zone 3	161	9	89	47	78	65	100	53
All sizes, climate zone 4	90	15	80	67	51	102	52	88
All sizes, climate zone 5	72	7	91	38	92	46	65	29
All climate zones, All sizes	125	73	94	260	74	355	78	363
All climate zones, Small			83	158	71	289		
All climate zones, Large			100	102	82	66		

ENERGY USE INTENSITY (EUI)—PER SF—MEASURED AT ONE DATA CENTER

Source: Large computer data center, Colorado Area, approx 30W/SF data equipment density, approx 40 W/SF total including A/C equipment, sub meter data.

Category	kBtu/SF
Data Center, computers and equipment only	865
Data Center, including computers, equipment, and cooling units	1170

CBECS CLIMATE ZONE MAP

Source: EIA/CBECS 2003

Heating Degree Days (HDD) are 65 degree Base

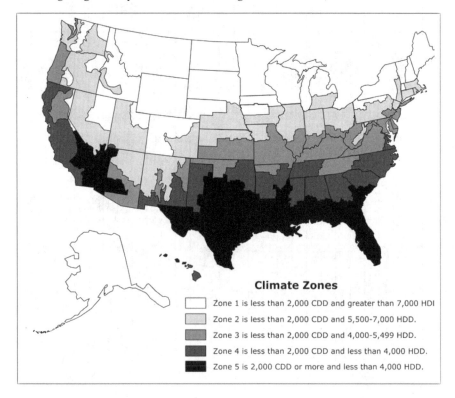

BUILDING USE CATEGORIES DEFINED (CBECS)
Source: EIA/CBECS 2003

Education
Buildings used for academic or technical classroom instruction, such as elementary, middle, or high schools, and classroom buildings on college or university campuses. Buildings on education campuses for which the main use is not classroom are included in the category relating to their use. For example, administration buildings are part of "Office," dormitories are "Lodging," and libraries are "Public Assembly."
- elementary or middle school
- high school
- college or university
- preschool or daycare
- adult education
- career or vocational training
- religious education

Food Sales
Buildings used for retail or wholesale of food.
- grocery store or food market
- gas station with a convenience store
- convenience store

Food Service
Buildings used for preparation and sale of food and beverages for consumption.
- fast food
- restaurant or cafeteria

Health Care—Inpatient
Buildings used as diagnostic and treatment facilities for inpatient care.
- hospital
- inpatient rehabilitation

Health Care—Outpatient
Buildings used as diagnostic and treatment facilities for outpatient care. Medical offices are included here if they use any type of diagnostic

medical equipment -if they do not, they are categorized as an office building.
- medical office clinic or other outpatient health care
- outpatient rehabilitation
- veterinarian

Laboratories

Buildings whose use is primarily scientific in nature, and that use a lot of specialized equipment. Computer labs, training facilities, and research and development buildings are considered "Office." A high school science or biology lab would most likely be part of a larger classroom building which would be considered "Education," but if it did happen to be a separate building, it would be classified as a lab. Lab departments within a hospital would be considered inpatient health care, but if it did happen to be a separate building, it would be classified as a lab.
- pharmaceutical labs
- water chemistry labs
- blood work labs
- test labs and other labs that are in separate buildings

Lodging

Buildings used to offer multiple accommodations for short-term or long-term residents
- motel or inn
- hotel
- dormitory, fraternity, or sorority
- convent or monastery
- shelter, orphanage, or children's home
- halfway house

Nursing
- retirement home
- nursing home, assisted living, or other residential care
- skilled nursing and other residential care buildings

Office

Buildings used for general office space, professional office, or administrative offices. Medical offices are included here if they do not use any type of diagnostic medical equipment (if they do, they are categorized

as an outpatient health care building).
- administrative or professional office
- government office
- mixed-use office
- bank or other financial institution
- medical office (see previous column)
- sales office
- contractor's office (e.g. construction, plumbing, HVAC)
- non-profit or social services
- research and development
- city hall or city center
- religious office
- call center

Retail—Other Than Mall
Buildings used for the sale and display of goods other than food.
retail store
- beer, wine, or liquor store
- rental center
- dealership or showroom for vehicles or boats
- studio/gallery

Public Assembly
Buildings in which people gather for social or recreational activities, whether in private or non-private meeting halls.
- social or meeting (e.g. community center, lodge, meeting hall, convention center, senior center)
- recreation (e.g. gymnasium, health club, bowling alley, ice rink, field house, indoor racquet sports)
- entertainment or culture (e.g. museum, theater, cinema, sports arena, casino, night club)
- library
- funeral home
- student activities center
- armory
- exhibition hall
- broadcasting studio
- transportation terminal

Public Order and Safety
Buildings used for the preservation of law and order or public safety.
- police station
- fire station
- jail, reformatory, or penitentiary
- courthouse or probation office

Religious Worship
Buildings in which people gather for religious activities, (such as chapels, churches, mosques, synagogues, and temples).

Service
Service Buildings in which some type of service is provided, other than food service or retail sales of goods
- vehicle service or vehicle repair shop
- vehicle storage/maintenance (car barn)
- repair shop
- dry cleaner or Laundromat
- post office or postal center
- car wash
- gas station
- photo processing shop
- beauty parlor or barber shop
- tanning salon
- copy center or printing shop
- kennel

Warehouse and Storage
Buildings used to store goods, manufactured products, merchandise, raw materials, or personal belongings (such as self-storage).
- refrigerated warehouse
- non-refrigerated warehouse
- distribution or shipping center

Appendix

OPERATING EXPENSES: PERCENT THAT ARE FROM UTILITY COSTS

Business Type	Percent	Source	Remarks
Colleges	3.2	American School and University (ASU) survey, 2007	Purchased utilities cost divided by Total operating expenses, including payroll (includes teaching staff payroll)
Day Care	2.1	U.S. Census Bureau, Business Expenses: 2002 (NAICS 6244)	Purchased utilities cost divided by Total operating expenses, including payroll
Healthcare - Hospitals	1.3	U.S. Census Bureau, Business Expenses: 2002 (NAICS 622)	Purchased utilities cost divided by Total operating expenses, including payroll
Healthcare - Outpatient	0.8	U.S. Census Bureau, Business Expenses: 2002 (NAICS 6214)	Purchased utilities cost divided by Total operating expenses, including payroll
Hotels and Motels	5.3	U.S. Census Bureau, Business Expenses: 2002 (NAICS 7211)	Purchased utilities cost divided by Total operating expenses, including payroll
K-12 Schools	2.1	American School and University (ASU) survey, 2007	Purchased utilities cost divided by Total operating expenses, including payroll (includes teaching staff payroll)
Linen and Uniform Supply	4.4	U.S. Census Bureau, Business Expenses: 2002 (NAICS 81233)	Purchased utilities cost divided by Total operating expenses, including payroll
Medical and Diagnostic Lab	0.6	U.S. Census Bureau, Business Expenses: 2002 (NAICS 6215)	Purchased utilities cost divided by Total operating expenses, including payroll
Museums, Historical Sites, and Similar Institutions	3.6	U.S. Census Bureau, Business Expenses: 2002 (NAICS 712)	Purchased utilities cost divided by Total operating expenses, including payroll
Nursing Care Facilities	2.1	U.S. Census Bureau, Business Expenses: 2002 (NAICS 6231)	Purchased utilities cost divided by Total operating expenses, including payroll

Business Type	Percent	Source	Remarks
Office Buildings – Leased Building - tenant salaries not included	20	"Experience Exchange Report", 2006, used with permission from BOMA International. All rights reserved.	Weighted average of Class A and Class B buildings that reported. Percentage value is based on $2.00/SF-yr cost for all utilities. Percentage includes variable and fixed expenses for the buildings, but does not include payroll expenses of the building tenants. This would be from the viewpoint of the building leasing entity, since the tenant salaries are not included.
Office Buildings - Including payroll (approximate)	1.25	BOMA figure for leased space is 20% of total O/M expenses less payroll, and $2.00/SF-yr. "Experience Exchange Report", 2006, used with permission from BOMA International. All rights reserved. If $2.00/SF-yr is 20% of total, the total is $10/SF-yr This figure was adjusted using an approximate value of $150/SF-yr for salaries in an office building, for a revised total cost of $160/SF-yr including payroll. The same $2/SF-yr for utilities is then 1.2% Source of salary burden approximation: Derived from gathered data in the following document: NEMI, National Energy Management Institute, Productivity Benefits Due to Improved Indoor Air Quality, August 1995, pp 4-8, 4-11.	Approximate value of purchased utilities cost divided by Total operating expenses, including payroll. See Chapter 8 – Building Operations and Maintenance **"Productivity Value"** for this and other categories of building use. "Office" value of $97/SF-yr was adjusted from 1995 dollars at 4% inflation.
Restaurant – Full Service	5.1	U.S. Census Bureau, Business Expenses: 2002 (NAICS 7221)	Purchased utilities cost divided by Total operating expenses, including payroll
Restaurant – Limited Service (Fast Food)	4.6	U.S. Census Bureau, Business Expenses: 2002 (NAICS 7222)	Purchased utilities cost divided by Total operating expenses, including payroll
Retail	2.7	U.S. Census Bureau, Business Expenses: 2002 (NAICS 44-45)	Purchased utilities cost divided by Total operating expenses, including payroll

Business Type	Percent	Source	Remarks
Supermarkets	4.9	U.S. Census Bureau, Business Expenses: 2002 (NAICS 445)	Purchased utilities cost divided by Total operating expenses, including payroll
Warehouse	4.9	U.S. Census Bureau, Business Expenses: 2002 (NAICS 493)	Purchased utilities cost divided by Total operating expenses, including payroll

ENERGY USE INTENSITY (EUI)—PER SF— FOR SOME MANUFACTURING OPERATIONS

Source: Author Experience, Colorado Area

Note: EUI in per-SF terms is often of little use in manufacturing audits because they are not benchmarked that way. A more useful metric would be kBtu per unit of production, per pound of steel, etc. This table serves to show the wide range and lack of pattern in manufacturing businesses.

General Description	Size	kBtu/SF-yr
Semiconductor Fabrication, 24x7	207 KSF	1622
Semiconductor Fabrication, 24x7.	579 KSF	1484
High Volume Bulk Jobber Electro-Plating, 1-1/2 shifts, all electric resistance heat.	4KSF	690
Hand tool forging and machining, 3 shifts	300 kSF	615
Bulk silicon crystal growing, electric resistance heating of salt vats.	27 KSF	612
Close Tolerance plastic injection molding, including some clean rooms.	62 KSF	479
Milk Product Processing / Packaging / Pasteurizing.	97 KSF	431
Light machining, annealing, Teflon coating and high temperature curing.	55 KSF	410
High Volume Electro-plating of unique shapes, 2-shifts.	27 KSF	291
Precision machining of small electronic components, some atmosphere-controlled sintering and vacuum metal deposition, 30% of area is a clean room. 2 shifts.	14 KSF	280
Metal Stamping, heat treating, plating, 2-1/2 shifts.	42 KSF	257
Light metal manufacturing and assembly, heat treating, plating, painting, welding, 2 shifts.	160 KSF	209
Light metal manufacturing and assembly, plating, painting, welding.	25 KSF	181
Bulk plastic injection molding.	12 KSF	161
Light assembly and testing of data storage systems. Large portion is offices and material storage, light energy use.	240 KSF	110

ENERGY USE INTENSITY (EUI)—IN PRODUCTION UNITS— SOME MFG. PROCESSES

Category		Source
Dairy, fluid milk processing	278 Btu per pound of milk	E Source "Food Products and Beverages Industry Snapshot Report" Aug 2003. Note: 51% of the energy use is in pasteurizing (heating and then cooling).
Meat processing	429 Btu per pound meat	E Source "Food Products and Beverages Industry Snapshot Report" Aug 2003. Note: 64% of the energy use is in refrigeration.
Breweries	516 Btu per pound of pasteurized malt	E Source "Food Products and Beverages Industry Snapshot Report" Aug 2003. Note: 51% of the energy use is in pasteurizing (heating and then cooling).
Ready Mix Concrete	384 Btu per pound of product	E Source Market Sector Snapshots, "Concrete Products", 2005. Note: 84% of energy use is from the mixing the concrete during transportation
Portland cement manufacturing, dry process	1474 Btu per pound of product	E Source Market Sector Snapshots, "Concrete Products", 2005. Note: 71% of energy use is from the kiln firing step

ASHRAE 90.1—ITEMS REQUIRED FOR ALL COMPLIANCE METHODS

Source: extracted from ASHRAE 90.1-2001, © American Society of Heating, Refrigerating and Air-Conditioning Engineers, Inc., www.ashrae.org.

Applies to all systems, regardless of whether the Prescriptive or Performance Method is used.

Chapter 11 of ASHRAE Standard 90.1 clarifies that in addition to meeting the energy budget (the point-based system); compliance requires that "all requirements of 5.2, 6.2, 7.2, 8.2, 9.2, and 10.2 are met." The "Mandatory Requirements" are listed in each chapter as follows:

5.2	Building Envelope	8.2	Power
6.2	HVAC	9.2	Lighting
7.2	Service Water Heating	10.2	Other Equipment

Some of the requirements are paraphrased. Extra *[author comments]* add clarity or explain the value.

Partial Listing of ASHRAE 90.1 Mandatory Requirements

These are in addition to ASHRAE 90.1 Energy Budget Point Requirements of Chapter 11 of the Standard. The standard is under continuous maintenance by ASHRAE committee; however these fundamental requirements are relatively unchanged and often overlooked by those using and enforcing the standard.

Section	Requirement
5.2 Building Envelope	Envelope Sealing: *[Buildings are expected to be constructed tightly. Before the work is covered up, ask to see all the sealing around each of these penetrations. Air plenums are common—these also need to be well sealed to prevent leakage]* The following areas of the building envelope shall be sealed, caulked, gasketed, or weather-stripped to minimize air leakage: • Joints around fenestration *[glass]* and door frames • Junctions between walls and foundations, between walls at building corners, between walls and structural floors or roofs, and between walls and roof or wall panels. • Openings at penetrations of utility services through roofs, walls, and floors

Section	Requirement
5.2 Building Envelope (Cont'd)	• Site-built fenestration *[glass]* and doors • Building assemblies used as ducts or plenums • Joints, seams, and penetrations of vapor retarders • All other openings in the building envelope
6.2 HVAC	**Equipment Efficiencies**: • Required mechanical equipment efficiency minimum standards are listed for all types of heating and cooling equipment. • Gas-fired and oil-fired force air furnaces with input ratings >= 225,000 Btuh cannot use standing pilot ignition, and must have either power venting or a flue damper. *[this provides assurance that the building is starting out with decent equipment efficiency]* **Load Calculations:** • Heating and cooling system design load calculations used for sizing systems and equipment are required. *[i.e. rules of thumb "per SF" average values, or other seat-of-the-pants methods are not acceptable—ask for a copy of the calculations]* **Controls:** • **Definition of "Zone":** For the following a "zone" is defined as "a space or group of spaces within a building with heating and cooling requirements that are sufficiently similar so that desired conditions (e.g. temperature can be maintained throughout using a single sensor (e.g. a single thermostat or temperature sensor) • **Zone Thermostatic Controls**: The supply of heating and cooling energy to each zone shall be individually controlled by thermostatic controls responding to temperatures with the zone. • All zone and loop controllers shall use control methodology that incorporates the application of control error reduction. (this requires either floating control or "integral" (DDC) control) • **Dead Band**: When used to control both heating and cooling, zone thermostatic control shall be capable of providing a deadband of at least 5 degrees F. Special occupancy or special applications are exempt, such as process control, nursing homes, data processing, museums, etc. • **Set Point Overlap Restriction**: For heating and cooling controls within a single zone, positive means (such as limit switches, mechanical stops, or software programming) shall be provided to prevent the heating set point from exceeding the cooling set point minus any proportional band. • **Off-Hour Controls**: HVAC systems having a design heating or cooling capacity greater than 65,000 Btu/h *[5.4 tons]* and a fan system power greater than 3/4 hp shall have all of the following off-hour controls: — Automatic Shutdown (time control, occupancy control, two-hour manual timer, or interlock with security system activation (unoccupied).

Appendix 381

Section	Requirement
HVAC (Cont'd)	— Set back Controls (55 degree or lower unoccupied heating setting) — Optimum Start Controls *[adjusts start-up time for equipment to be at-temperature just in time for occupancy]* • **Ventilation and Zone Air Flow System Controls:** — HVAC systems serving zones that are intended to operate or be occupied non-simultaneously shall be divided into isolation areas. Zones may be grouped into a single isolation areas provided it does not exceed 25,000 SF of conditioned floor areas nor include more than one floor. — Each isolation area shall be equipped with independently controlled isolation devices *[motorized dampers]* capable of automatically shutting off the supply of conditioned air and outside air to and exhaust air from the area. For central systems and plants, controls and devices shall be provided to allow stable system and equipment operation for any length of time while serving only the smallest isolation areas served by the system or plant. *[The previous two items will give larger buildings the ability to be cordoned off and heated/cooled in sections if desired for energy savings.]* — All outdoor air supply and exhaust hoods, vents, and ventilators that serve conditioned spaces shall be equipped with motorized dampers that will automatically shut when the spaces served are not in use—not required if outside air intake or exhaust is 300 CFM or less. *[These are minor air flows.]* — Maximum damper leakage for these isolation dampers is given in CFM per SF of damper size, at 1 in. w.c. pressure. *[The leakage spec assures a reasonable quality damper.]* • Ventilation Controls for High-Occupancy Areas: Systems with outside air capacities greater than 3000 CFM serving areas having an average design occupancy density exceeding 100 people per 1000 SF shall include means to automatically reduce outside air intake below design rates when spaces are partially occupied. Ref ASHRAE 62 and local standards. Not required if acceptable heat recovery is used to pre-condition the raw outside air. **Duct Sealing:** • Seal levels are prescribed in a table, including seal requirements for return air plenums. **As-Built Drawings:** • Required to be accurate, and delivered within 90 days after the date of system acceptance. **O/M Manuals:** • For all equipment and systems provided that require maintenance, the manuals must include: — Submittal data — Required routine maintenance

Section	Requirement
:	— HVAC controls system maintenance and calibration information, wiring diagrams, schematics, and control sequence descriptions. — A complete narrative of how each system is intended to operate, including suggested set points. **System Balancing:** — Required, with written report, for all HVAC systems serving zones with a total conditioned areas seceding 5000 SF. — Method used for balancing shall be "proportional balancing" which reduces throttling losses. — For fans greater than 1 hp, fan speed shall be adjusted to meet design flow conditions. *[This means changing pulleys.]* — For pumps over 10 hp, the impeller shall be trimmed if throttling to meet design flows conditions produces more than 5 percent of the nameplate hp. **Commissioning** • HVAC control systems shall be tested to ensure that control elements are calibrated, adjusted, and in proper working condition. For projects larger than 50,000 SF (except warehouses and semi-heated spaces), detailed instructions for commissioning HVAC systems shall be provided by the designer in plans and specifications. *[Commissioning is verification that you are getting what you asked for and paid for.]*
7.2	**Domestic Hot Water:**
Service Water Heating	• Hot Water Recirculating or Heat Trace Systems: Controls must automatically turn off the pipe temperature maintenance system during extended periods when hot water is not required. • Maximum Temperature to Lavatory Faucets: 110 degrees F. • Heat traps are required on storage water heaters and storage tanks that don't have recirculating systems. [reduces standby losses from thermal siphoning]
Pools:	• Pool heaters shall not use standing pilots. • Heated pools shall be equipped with a pool cover. Pools heated to more than 90 degrees F shall have a pool cover with a minimum insulation value of R-12. Exception to this is if at least 60 percent heated from site recovered energy or solar energy.
8.2	**As-Built Drawings:**
Power	• Required to be accurate, and delivered within 30 days after the date of system acceptance.

Section	Requirement
	O/M Manuals: • For all equipment and systems provided that require maintenance, the manuals must include: — Submittal data — Required routine maintenance — A complete narrative of how each system is intended to operate.
9.2 Lighting	**Automatic Interior Lighting Shutoff:** • Interior lighting for buildings 5000 SF and larger shall be controlled with an automatic control device to shut off building lighting in all spaces, either • Time scheduling device • Occupancy Sensor • Signal from another system indicating the area is unoccupied. **Space Control:** • There must be individual controls for the general lighting in each zone, with separate controls for each zone of certain size (2500 SF per zone for a space of 10,000 SF, and a maximum of 10,000 SF per zone for larger spaces.) • Where space controls override the master automatic shutoff, the user override cannot exceed four hours. **Outdoor Lighting Efficiency:** • All outside lighting must be of a minimum efficiency (60 lumens per watt), unless required for health or safety. [this means no incandescent lighting can be used outside] **Outdoor Lighting Control:** • All outdoor lighting, except where lighting is used for security, safety, etc., shall be automatically turned off either by a photo cell or an astronomical time switch, when the lighting is not required. • The amount of interior lighting "Watts per Square Foot" are limited by usage, and form a Lighting Power Budget the designers need to stay within. [Ask for calculations to show that the watts per SF does not exceed the Energy Code Lighting Budget—lighting is a very large component of energy use in most buildings]
10.2 Other Equipment	**Electric Motor Efficiency:** • Minimum efficiency of electric motors is defined in a table. Basis is EPACT-1992 (U.S. Energy Policy Act)

TOP 15 EMERGING TECHNOLOGIES—2002 (DOE)

Source: Energy Consumption Characteristics of Commercial Building HVAC Systems Volume III: Energy Savings Potential, 2002, Building Technologies Program, DOE.

Technology Option	Technology Status	Technical Energy Savings Potential (quads)
Adaptive/Fuzzy Logic Controls	New	0.23
Dedicated Outdoor Air Systems	Current	0.45
Displacement Ventilation	Current	0.20
Electronically Commutated Permanent Magnet Motors	Current	0.15
Enthalpy/Energy Recovery Heat Exchangers for Ventilation	Current	0.55
Heat Pumps for Cold Climates (Zero-Degree Heat Pump)	Advanced	0.1
Improved Duct Sealing	Current/New	0.23
Liquid Desiccant Air Conditioners	Advanced	0.2 / 0.06[1]
Microchannel Heat Exchanger	New	0.11
Microenvironments / Occupancy-Based Control	Current	0.07
Novel Cool Storage	Current	0.2/ 0.03[2]
Radiant Ceiling Cooling / Chilled Beam	Current	0.6
Smaller Centrifugal Compressors	Advanced	0.15
System/Component Diagnostics	New	0.45
Variable Refrigerant Volume/Flow	Current	0.3

Summary of the 15 Technology Options Selected for Refined Study

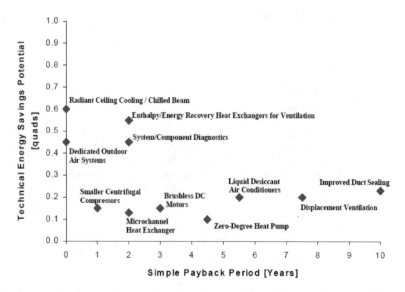

Estimated Technical Energy Savings Potential and Simple Payback Periods for the 15 Options

Appendix

HVAC RETROFITS FOR THE THREE WORST SYSTEMS
- Constant Volume Reheat
- Multi-zone
- Double Duct

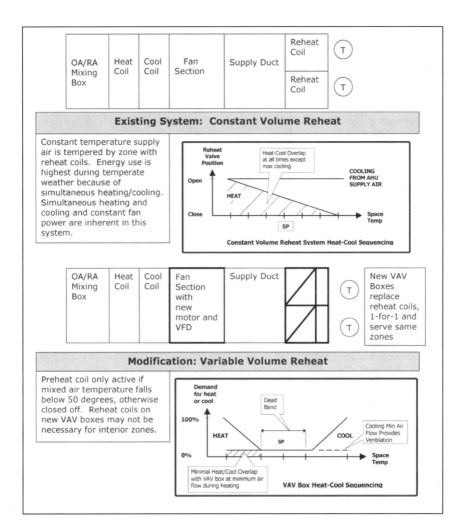

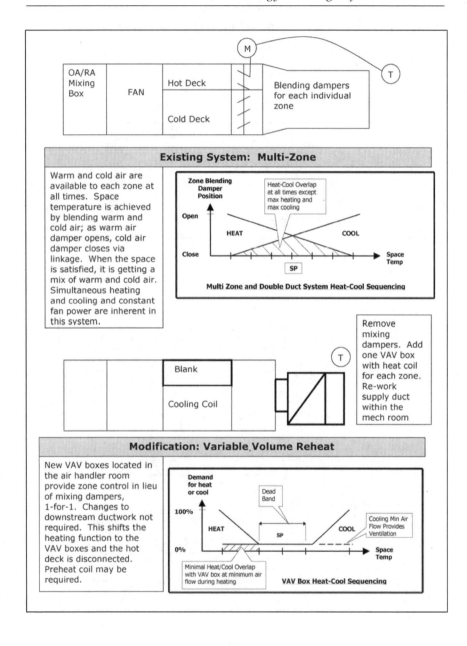

Appendix

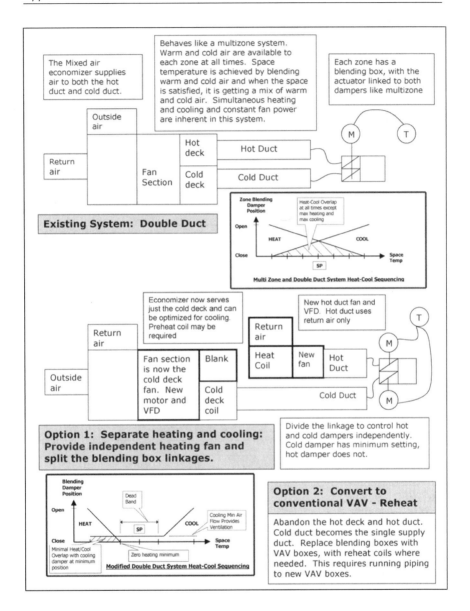

SERVICE LIFE OF VARIOUS SYSTEM COMPONENTS

Note: The figures given are the end of life period when the equipment is scrapped. This does not mean there are no repairs along the way. For example, it is almost certain that an air conditioner or heat pump will get several repairs including a new compressor during the advertised 'life'.

Equipment	Normal Expected Replacement Life	Source
Hot Water Boiler	25 years	1
Steam Boiler	30 years	1
Steam Traps	7 years	3
Conventional Direct Gas-Fired Tank-Type Domestic Water Heater	8-12 years	4
Heat Pump Water Heater	10 years	3
Solar Water Heater	15 years	3
Centrifugal Chillers	23 years	1
Reciprocating Chiller	12-14 years	2
Screw Chiller	20 years	3
Galvanized Cooling Towers	20 years	1
Package Rooftop A/C Unit	15 years	1
Water Cooled Package Unit	15 years	1
Split System A/C	15 years	1
Fan Coil	20 years	2
VAV Boxes	20 years	1
Hot Water Unit Heaters	20 years	1
Electric Unit Heaters	13	1
PTAC (Packaged Terminal Air Conditioner)	10-15 years	2,1
Computer Room Air Conditioner	10-15 years	2,3
Gas Furnace	18 years	1
Gas Fired Radiant Tube Heater	10 years	3
Air Source Heat Pump	15 years	1
Water Source or Ground Source Heat Pump (closed loop)	19 years	1
Ground Source Heat Pump Bore Field (HDPE pipe material life is 50 years. System life may be less and may be limited by the bond between the grout, the pipe, and the earth and the degradation of heat transfer interface to the earth)	30+ years (estimated)	5

Appendix

Equipment	Normal Expected Replacement Life	Source
Indoor Air Handler	20-25 years	2
Air-Side Economizers	10 years	3
Water-Side Economizers	11 years	3
Electric Baseboard Heat	10-15 years	4
Electric Duct Heater	15 years	1
Hot Water Baseboard Heat	25 years	1
Base Mounted Pump	20 years	1
Sump Pump	10 years	1
Utility Fans	20 years	2
Ductwork (metal)	30 years	1
Air Curtain	10 years	3
Polyethylene Strip Curtain	3 years	3
Kitchen Exhaust Hood Make-Up Air Tempering Unit	10 years	3
Shell and Tube Heat Exchanger	24 years	1
Heat Pipe Heat Recovery	14 years	3
Rotary Wheel Heat Recovery	11 years	3
Heat Recovery from Refrigeration Condensers	11 years	3
Thermal Energy Storage System (TES) - Ice	19 years	3
Thermal Energy Storage System (TES) - Water	20 years	3
Direct Evaporative Cooling	7-10 years	5
Evaporative Pre-Cooling	8-12 years	5
Indirect-Direct Evaporative Cooling	15-20 years	5
Evaporative Cooling Cellulose Media	5 years	6
Evaporative Cooling Felt Pads	2 years	6
Air Washer	17 years	1
Motors	15-17 years	3
VFD	15 years	3
Motor Starter	17 years	3
Air Compressor	20	2
Lighting Fixture	20 years	3
Ballast – all types	12 years	3
Motion Sensor	10 years	3
Dimming Systems	20	3

Equipment	Normal Expected Replacement Life	Source
Double Pane Windows	12-20 years	6
Solar Shade Film	7-10 years	3
Molded Insulation	20 years	1
Blanket Insulation	24 years	1
Control Valves	20 years	1
Dampers	20 years	1
Valve/Damper Actuator - pneumatic	20 years	1
Valve/Damper Actuator – hydraulic	15 years	1
Valve/Damper Actuator – mini hydraulic (for terminal units)	5 years	7
Valve/Damper Actuator – electric – oil filled	10-15 years	7
Valve/Damper Actuator – electric – open air	5-7 years	7
Valve/Damper Actuator – self contained (system powered)	10 years	1
Valve/Damper Actuator – Residential style "clock motor" terminal valves	5 years	7
"Active" control sensors and transmitters (powered-type)	5 years	7
Pneumatic Controls – General	20 years	1
Analog Electronic Controls - General	7-10 years	7
DDC Controls (before made obsolete by technology advances)	7-10 years	7

Sources:
1. ASHRAE Applications Handbook, 2003. © American Society of Heating, Refrigerating and Air-Conditioning Engineers, Inc., www.ashrae.org.
2. Iowa Department of Natural Resources, 2002, "Life Cycle Cost Analysis Guidelines – 2002"
3. "Service Life of Energy Conservation Measures" Bonneville Power Administration (July 14, 1987)
4. Stoffer Inspections, L.C. Home Page, "Life Cycles and Approximate Costs to Repair/Replace/Upgrade", 2004 (home inspector)
5. Author estimate
6. Manufacturer's Literature
7. Author's field experience

EQUATING ENERGY SAVINGS TO PROFIT INCREASE

Energy is an operating expense and energy savings go to the bottom line, providing an increase in overall profitability. The following tables illustrate this effect. The largest profit boost will occur when the businesses profit margin is small and when the percent of operating expense that is from energy is large.

Profit Increase from Energy Savings

5% Energy Savings
Table shows revised profit value

Original Profit Margin	Energy Cost % of Total Operating Cost									
	1%	2%	3%	4%	5%	6%	7%	8%	9%	10%
1%	1.1%	1.1%	1.2%	1.2%	1.3%	1.3%	1.4%	1.4%	1.5%	1.5%
2%	2.1%	2.1%	2.2%	2.2%	2.3%	2.3%	2.4%	2.4%	2.5%	2.5%
3%	3.1%	3.1%	3.2%	3.2%	3.3%	3.3%	3.4%	3.4%	3.5%	3.5%
5%	5.1%	5.1%	5.2%	5.2%	5.3%	5.3%	5.4%	5.4%	5.5%	5.5%
10%	10.1%	10.1%	10.2%	10.2%	10.3%	10.3%	10.4%	10.4%	10.5%	10.6%
20%	20.1%	20.1%	20.2%	20.2%	20.3%	20.4%	20.4%	20.5%	20.5%	20.6%
30%	30.1%	30.1%	30.2%	30.3%	30.3%	30.4%	30.5%	30.5%	30.6%	30.7%

10% Energy Savings
Table shows revised profit value

Original Profit Margin	Energy Cost % of Total Operating Cost									
	1%	2%	3%	4%	5%	6%	7%	8%	9%	10%
1%	1.1%	1.2%	1.3%	1.4%	1.5%	1.6%	1.7%	1.8%	1.9%	2.0%
2%	2.1%	2.2%	2.3%	2.4%	2.5%	2.6%	2.7%	2.8%	2.9%	3.0%
3%	3.1%	3.2%	3.3%	3.4%	3.5%	3.6%	3.7%	3.8%	3.9%	4.0%
5%	5.1%	5.2%	5.3%	5.4%	5.5%	5.6%	5.7%	5.8%	6.0%	6.1%
10%	10.1%	10.2%	10.3%	10.4%	10.6%	10.7%	10.8%	10.9%	11.0%	11.1%
20%	20.1%	20.2%	20.4%	20.5%	20.6%	20.7%	20.8%	21.0%	21.1%	21.2%
30%	30.1%	30.3%	30.4%	30.5%	30.7%	30.8%	30.9%	31.0%	31.2%	31.3%

15% Energy Savings
Table shows revised profit value

Original Profit Margin	Energy Cost % of Total Operating Cost									
	1%	2%	3%	4%	5%	6%	7%	8%	9%	10%
1%	1.2%	1.3%	1.5%	1.6%	1.8%	1.9%	2.1%	2.2%	2.4%	2.5%
2%	2.2%	2.3%	2.5%	2.6%	2.8%	2.9%	3.1%	3.2%	3.4%	3.6%
3%	3.2%	3.3%	3.5%	3.6%	3.8%	3.9%	4.1%	4.3%	4.4%	4.6%
5%	5.2%	5.3%	5.5%	5.6%	5.8%	6.0%	6.1%	6.3%	6.4%	6.6%
10%	10.2%	10.3%	10.5%	10.7%	10.8%	11.0%	11.2%	11.3%	11.5%	11.7%
20%	20.2%	20.4%	20.5%	20.7%	20.9%	21.1%	21.3%	21.5%	21.6%	21.8%
30%	30.2%	30.4%	30.6%	30.8%	31.0%	31.2%	31.4%	31.6%	31.8%	32.0%

Profit Increase from Energy Savings

20% Energy Savings
Table shows revised profit value

Original Profit Margin	Energy Cost % of Total Operating Cost									
	1%	2%	3%	4%	5%	6%	7%	8%	9%	10%
1%	1.2%	1.4%	1.6%	1.8%	2.0%	2.2%	2.4%	2.6%	2.9%	3.1%
2%	2.2%	2.4%	2.6%	2.8%	3.0%	3.2%	3.4%	3.7%	3.9%	4.1%
3%	3.2%	3.4%	3.6%	3.8%	4.0%	4.3%	4.5%	4.7%	4.9%	5.1%
5%	5.2%	5.4%	5.6%	5.8%	6.1%	6.3%	6.5%	6.7%	6.9%	7.1%
10%	10.2%	10.4%	10.7%	10.9%	11.1%	11.3%	11.6%	11.8%	12.0%	12.2%
20%	20.2%	20.5%	20.7%	21.0%	21.2%	21.5%	21.7%	22.0%	22.2%	22.4%
30%	30.3%	30.5%	30.8%	31.0%	31.3%	31.6%	31.8%	32.1%	32.4%	32.7%

25% Energy Savings
Table shows revised profit value

Original Profit Margin	Energy Cost % of Total Operating Cost									
	1%	2%	3%	4%	5%	6%	7%	8%	9%	10%
1%	1.3%	1.5%	1.8%	2.0%	2.3%	2.5%	2.8%	3.1%	3.3%	3.6%
2%	2.3%	2.5%	2.8%	3.0%	3.3%	3.6%	3.8%	4.1%	4.3%	4.6%
3%	3.3%	3.5%	3.8%	4.0%	4.3%	4.6%	4.8%	5.1%	5.4%	5.6%
5%	5.3%	5.5%	5.8%	6.1%	6.3%	6.6%	6.9%	7.1%	7.4%	7.7%
10%	10.3%	10.6%	10.8%	11.1%	11.4%	11.7%	12.0%	12.2%	12.5%	12.8%
20%	20.3%	20.6%	20.9%	21.2%	21.5%	21.8%	22.1%	22.4%	22.8%	23.1%
30%	30.3%	30.7%	31.0%	31.3%	31.6%	32.0%	32.3%	32.7%	33.0%	33.3%

30% Energy Savings
Table shows revised profit value

Original Profit Margin	Energy Cost % of Total Operating Cost									
	1%	2%	3%	4%	5%	6%	7%	8%	9%	10%
1%	1.3%	1.6%	1.9%	2.2%	2.5%	2.9%	3.2%	3.5%	3.8%	4.1%
2%	2.3%	2.6%	2.9%	3.2%	3.6%	3.9%	4.2%	4.5%	4.8%	5.2%
3%	3.3%	3.6%	3.9%	4.3%	4.6%	4.9%	5.2%	5.5%	5.9%	6.2%
5%	5.3%	5.6%	6.0%	6.3%	6.6%	6.9%	7.3%	7.6%	7.9%	8.2%
10%	10.3%	10.7%	11.0%	11.3%	11.7%	12.0%	12.4%	12.7%	13.1%	13.4%
20%	20.4%	20.7%	21.1%	21.5%	21.8%	22.2%	22.6%	23.0%	23.3%	23.7%
30%	30.4%	30.8%	31.2%	31.6%	32.0%	32.4%	32.8%	33.2%	33.6%	34.0%

Appendix

Profit Increase from Energy Savings

35% Energy Savings
Table shows revised profit value

Original Profit Margin	Energy Cost % of Total Operating Cost								
	1%	2%	3%	4%	5%	6%	7%	8%	9%	10%
1%	1.4%	1.7%	2.1%	2.4%	2.8%	3.2%	3.5%	3.9%	4.3%	4.7%
2%	2.4%	2.7%	3.1%	3.4%	3.8%	4.2%	4.6%	4.9%	5.3%	5.7%
3%	3.4%	3.7%	4.1%	4.5%	4.8%	5.2%	5.6%	6.0%	6.4%	6.7%
5%	5.4%	5.7%	6.1%	6.5%	6.9%	7.3%	7.6%	8.0%	8.4%	8.8%
10%	10.4%	10.8%	11.2%	11.6%	12.0%	12.4%	12.8%	13.2%	13.6%	14.0%
20%	20.4%	20.8%	21.3%	21.7%	22.1%	22.6%	23.0%	23.5%	23.9%	24.4%
30%	30.5%	30.9%	31.4%	31.8%	32.3%	32.8%	33.3%	33.7%	34.2%	34.7%

40% Energy Savings
Table shows revised profit value

Original Profit Margin	Energy Cost % of Total Operating Cost								
	1%	2%	3%	4%	5%	6%	7%	8%	9%	10%
1%	1.4%	1.8%	2.2%	2.6%	3.1%	3.5%	3.9%	4.3%	4.8%	5.2%
2%	2.4%	2.8%	3.2%	3.7%	4.1%	4.5%	4.9%	5.4%	5.8%	6.3%
3%	3.4%	3.8%	4.3%	4.7%	5.1%	5.5%	6.0%	6.4%	6.8%	7.3%
5%	5.4%	5.8%	6.3%	6.7%	7.1%	7.6%	8.0%	8.5%	8.9%	9.4%
10%	10.4%	10.9%	11.3%	11.8%	12.2%	12.7%	13.2%	13.6%	14.1%	14.6%
20%	20.5%	21.0%	21.5%	22.0%	22.4%	23.0%	23.5%	24.0%	24.5%	25.0%
30%	30.5%	31.0%	31.6%	32.1%	32.7%	33.2%	33.7%	34.3%	34.9%	35.4%

50% Energy Savings
Table shows revised profit value

Original Profit Margin	Energy Cost % of Total Operating Cost								
	1%	2%	3%	4%	5%	6%	7%	8%	9%	10%
1%	1.5%	2.0%	2.5%	3.1%	3.6%	4.1%	4.7%	5.2%	5.8%	6.3%
2%	2.5%	3.0%	3.6%	4.1%	4.6%	5.2%	5.7%	6.3%	6.8%	7.4%
3%	3.5%	4.0%	4.6%	5.1%	5.6%	6.2%	6.7%	7.3%	7.9%	8.4%
5%	5.5%	6.1%	6.6%	7.1%	7.7%	8.2%	8.8%	9.4%	9.9%	10.5%
10%	10.6%	11.1%	11.7%	12.2%	12.8%	13.4%	14.0%	14.6%	15.2%	15.8%
20%	20.6%	21.2%	21.8%	22.4%	23.1%	23.7%	24.4%	25.0%	25.7%	26.3%
30%	30.7%	31.3%	32.0%	32.7%	33.3%	34.0%	34.7%	35.4%	36.1%	36.8%

INTEGRATED DESIGN EXAMPLES

An important concept in economically successful energy efficiency designs is that of Integrated Design, whereby energy reduction in one element results in a reduction in size of a related supporting element, thereby subsidizing or even paying for the measure. Examples are:
- The added cost of high performance suspended film windows can be offset in some climates by the elimination of perimeter fin-tube heating.
- Down-sizing cooling systems in conjunction with reduced lighting power design.
- Down-sizing electrical systems in conjunction with reduced HVAC equipment load and sizes.
- Down-sizing electrical systems in conjunction with reduced transport energy costs (smaller fans and pumps from amply sized ducts and pipes).
- Down-sizing cooling systems in conjunction with improved window shading coefficients, films, or exterior shading systems.
- Down-sizing heating and cooling systems by using 1% or 2% ASHRAE design outdoor weather conditions instead of 0.4%, allowing the temperature to drift up a few degrees a few hours of the year.
- Down-sizing a boiler or hot water unit by virtue of selecting higher efficiency equipment. Since output is the design driver, it is often possible to utilize the next smaller size unit, but at higher efficiency, to achieve the same result.
- Down-sizing primary heating cooling equipment in conjunction with upgrades in envelope elements like insulation or, especially, window shading.
- Down-sizing primary heating and cooling equipment in conjunction with heat recovery systems.
- Down-sizing fan and pump motors, and primary cooling equipment, by increasing duct and pipe sizes, filter areas, coil areas, etc., as the tradeoff for using less transport energy. Note that the extra heat of transport energy elements, especially fans, often drives the equipment size up a notch.
- Down-sizing overhead lighting and HVAC cooling via an owner commitment to a greater use of task lighting.

ENERGY AUDIT APPROACH FOR COMMERCIAL BUILDINGS

Note: These are the basics. Depth of data gathering, analysis, and detail will vary.

Preliminary
- Identify the business principal building activity, and understand basic common drivers of energy use.
- Know the top 3-5 end uses that probably constitute the bulk of energy use.

Utility Rate Review
- Identify unit cost of energy for gas and electric, and on and off peak times.
- Identify unit cost of demand kW, as well as any ratchet clause provisions.

Utility History Review
- Identify patterns and anomalies. These may be unexplained high or low uses associated with particular months of the year.
- Establish the energy use index (kBtu/SF-yr) and compare to benchmark data if available.
- Benchmark data will gage whether the customer's use seems reasonable, or unusually high or low.
- Benchmark data will also provide early indication of relative potential savings. If the benchmark is valid, then a customer with energy use per SF above that can easily see what percent of reduction is needed to fall in line with their peers.
- For buildings with multiple 'business" use, determine proportions and create an "adjusted" benchmark. This will provide insight as to where the energy use should be
- In most cases, use kBtu/SF-yr as the unit, not $/SF-yr since this is more stable (utility rates change).
- Determine load profile by month for insight into end use patterns. For example, boilers left running all summer are usually readily apparent in such graphs if the only other source of gas use is domestic hot water.
- If available obtain load profile by hour for even better insight. Few customers have time of use meters, but it is good to ask.
- Determine load factor and fraction of bill that is demand. Poor load factors are sometimes an opportunity to spread out load.

- Review energy use with respect to weather to understand the extent to which energy use is weather dependent.
- Determine the overall 'blended' rate of electric cost in $/kW and $/therm. Note that the blended rate is applicable to buildings that do not have demand charges, or where demand charges are not calculated separately. If demand savings are to be calculated, the blended rate cannot be used.

Questionnaire (See <u>Appendix</u> for Sample Questionnaire)
- Inquire about usage habits, hours of occupancy, number of people, when equipment is turned on and off, etc.
- Inquire about basic energy using systems including lights, HVAC, process equipment.
- Inquire about computer rooms, size, and cooling load. Cooling load is a good indicator of computer equipment actual load, since they balance.
- Ask questions that lead to the understand of the primary energy use points.
- Determine if process equipment dominates the usage.
- Prepare the customer to have a person to escort you through the facility for 4-8 hours Large facilities will take more time.

Analysis and Report
- Estimate energy use from various sources. Compare sum total of expected energy use to actual utility bills and typical end-use break down where applicable benchmarks and "pie charts" exist. This step is a vital sanity check.
- Take great care not to over-estimate savings. Not only would owner confidence be lost, but their budget may depend upon the accuracy of the estimates.
- Estimate demand savings, but only if the demand savings will occur during a peak time.
- Look hard at control routines being used.
- Look at ventilation. Often ventilation is LESS than it is supposed to be and correcting this will ADD energy costs.
- For equipment side, larger and less efficiency equipment should be targeted.
- Recommendations should begin with points of use and work back
- Group the recommendations.

- Strategic—smart choices to make for long term energy frugality.
- Low-Cost—relatively easy, quick items.
- Capital—investment required, cost identified, payback or ROI
- If costs are estimated, include design fees, escalation, contingency, and general conditions. Remember that retrofit costs are higher than new construction costs.
- For equipment that is in need of replacing anyway, the "cost" should not be normal replacement, it should be only for high efficiency upgrade. Energy Conservation payback should not be saddled with normal replacement costs. These should be budgeted anyway for normal expected life spans.
- Identify system issues noted, safety issues noted, etc. that provide additional value to the customer.
- Non-energy benefits, such as labor savings and pollution avoidance, may or may not be included.

ENERGY AUDIT LOOK-FOR ITEMS

General
- Visit major equipment and large points of energy use
- Walk the interior spaces and spot-measure temperatures and lighting levels
- If utility graphs show usage is temperature dependent, look hard at HVAC system
- Look for large exhaust flow points that are adjacent to coincident large make-up air points
- Look for negative pressure building

"Look For" Items—Maintenance
- Systems well understood?
- Mechanical areas kept clean?
- Heat exchangers kept clean?
- Chiller and boiler heads been off lately?
- How does the cooling tower basin water look? That's what the condenser tubes see
- Look at the filters and also the coils inside—often the filters are changed just before the audit starts

"Look For" Items—Point Of Use Conservation

- Over-lit areas
- Areas kept too warm or too cold. ASHRAE Comfort Envelope indicates 75 deg F dry bulb/50 percent rH in summer and 70 deg F dry bulb winter will satisfy most of the occupants most of the time, if appropriately dressed.

Source: ASHRAE Standard 55, 2004, © American Society of Heating, Refrigerating and Air-Conditioning Engineers, Inc., www.ashrae.org.

So, cooling below 75 deg F and/or heating above 70 deg F is an opportunity.

- Lights on in unoccupied areas, and also outdoor lights on during the day
- Heat/cool overlap
- Control valves and dampers that don't fully close (leaking hot water control valves can account for up to 20 percent of summer cooling load if the boilers are on!)
- Kitchen hoods left on all the time
- For process equipment, determine the relationship between production and energy use. E.g. when equipment is idle, is it still consuming energy?
- Large transformers located indoors (heat load)
- Small 24-7 cooling loads served by large house systems, forcing them to run continuously for this small area
- Computer rooms. These are very large energy users per SF. Energy Use index can be 10x that of other areas, per SF.
- High flow plumbing fixtures, esp. where hot water is used. Restricted flow fixtures, low volume basins or tubs, etc. can be effective to nip this energy and water use in the bud.

"Look For" Items—Automatic Temperature Control

- Continuous operation of equipment
- Duct static pressure set too high—try 1 inch w.c. and see if it works OK
- Supply air reset on VAV air handlers
- Space temperature adjustments allowed from local tenant
- Chiller condenser water temperature settings too high
- Control overlap in air handlers for sequenced items
- Chillers left on in winter

- Boilers left on in summer
- VAV minimums set too high
- VAV box minimum air flow setting for heating is higher than for minimum cooling.
- If trends and logs are available, obtain run hours and pct load for large equipment. This will establish load profiles.

"Look For" Items—Demand Charges
- Equipment run concurrently that could be spread out
- Electric space heating that could be staged through the control system
- Over-sized equipment

"Look For" items—Envelope
- Single-pane glass
- No wall insulation
- Minimal (1-2 inch) roof insulation. A good place to view this is the cut-out at a roof scuttle.
- Leaks around doors, window frames, and openable windows
- Large east and west glass exposures that aren't shaded or low-E coated.
- Return plenum not sealed at building perimeter, coupled to the envelope.

"Look For" items—Equipment Efficiency
- Electric resistance heat. Gas usually costs less, for simple heating tasks.
- Inefficient lighting (incandescent, T-12 fluorescent, HID)
- HVAC air economizer and compressor lock out below 55 deg F
- Old equipment that may be inefficient (HVAC, inverters). Spec grade equipment more than 10 years old can consume 50 percent more power than new high efficiency A/C units.
- HVAC fan horsepower in relation to cooling horsepower. Common (but not great) practice ends up with fan HP being half of the cooling HP, or 1/3 of the total. This is usually indicative of ductwork downsized to save initial dollars, but this builds-in excess utility costs for the life of the building.
- Equipment in poor condition—especially dirty or fouled heat exchange surfaces, or loose fins on air-cooled equipment. Measure approach temperatures to verify. Usually an approach over 5 degrees

indicates fouling for air coils and shell/tube heat exchangers. For boilers, the flue temperature should not be more than 100 degrees above the leaving fluid or steam temperature if operating normally.
- Boilers with excess air (combustion test required).
- Cooling towers undersized in proportion to equipment served. Rules of thumb for good energy efficiency from amply sized cooling towers are 0.05 kW/ton and 7 degree approach.
- Standby losses from idling equipment
- Domestic water heater "side arm" from the boiler, requiring the boiler to be active year-round
- Short cycling of primary equipment during part load demand creates standby losses for gas fired equipment

Appendix 401

ENERGY AUDIT—SAMPLE QUESTIONNAIRE

PRE-AUDIT CUSTOMER QUESTIONAIRRE—
NOTE N/A FOR SECTIONS THAT DO NOT APPLY.
NOTE: This form serves to gather information for the audit work, and to raise awareness of the customer on the subject of energy efficiency.

Customer Objectives

What is the desired outcome and benefit of this audit?	
What energy savings projects/programs have you implemented in the past 24 months?	
What energy savings projects/programs are currently planned?	
Will your budget for the next 12 months allow for investment in energy savings programs?	
If yes, what payback period or internal rate of return (IRR) do you hope to achieve?	

Documentation Requested

If you have them, please provide the following (copies) for us to take with us and review: • Automatic control shop drawings and sequences of operation. • This would include DDC and conventional controls (pneumatic, etc)	
If you have them, please have ready the following documents for review while on site: • Architectural drawings • Mechanical design drawings. • Mechanical one-line piping diagrams if distributed chilled water and hot water systems are used. • Mechanical equipment schedules that show air handlers, outside air quantities, fan and pump horsepower, Btuh capacity for cooling and heating, VAV box max and min settings, chillers, boilers, cooling towers, and other similar equipment. • Electrical design drawings and one-line power distribution diagrams. • Electrical lighting schedules • Test and balance reports • Submittal data • O/M data	

General

Can you provide a facility map showing the different areas, that we can have to write on? This could be an 'evacuation' or 'life safety' 11x17 map.	
Overall building square footage.	
Percent that the building is occupied/utilized. Any areas closed off?	
Indicate the building SF that is heated and cooled.	
Indicate the building SF that is heated and cooled by the primary equipment (chillers and boilers) vs. unitary packaged HVAC equipment.	
What is the Building Vintage? • Different ages of different additions? • Please describe each major addition, when it occurred, and what was done.	
What are the primary activities in the building? For each use that applies, indicate the name and areas in SF. • Assembly • Data Center • Education • Food Service • Health Care—Inpatient • Health Care—Outpatient • Laboratory • Laundry • Library/Museum • Lodging • Manufacturing • Multi-family • Office • Religious Worship • Retail • Service • Warehouse and Storage • Other	
Number of people that normally occupy the building, their hours of occupancy.	
Any recent improvements to the building that affect energy use?	
Any planned improvements to the building that would affect energy use?	

Appendix

Utilities

Is energy use monitored, tracked, logged, and benchmarked by month and year including usage and demand?	
Are there any measures in place to monitor and control building electric demand? If monthly or annual usage or demand increases past the benchmark, what happens?	
If any areas are sub-metered, a 12-month history of electrical or other energy use as applicable would be helpful. Example would be leased areas that happen to be sub-metered. If not metered separately, ignore this item.	

Envelope

Has any insulation been added or replaced as part of a roofing project since the building was built?	
Is the building glass single or double pane?	
Are there areas where cold draft complaints exist, to indicate loose or leaky construction? If yes, which areas are these?	
Do you notice air pushing out the front entrance doors or air being sucked in at the front entrance doors? Any other places where air movements are noticeable, such as rear doors, elevators, or corridors?	
Is there a building entrance vestibule?	
Any cold areas over ceilings near the edges of the building in winter?	

Lighting

What types of lighting are used and where? • Fluorescent T-8 • Fluorescent T-12 • Compact fluorescent • Metal Halide or other HID lights • Incandescent or halogen lights	
Are any areas known to be over-lit? Other lighting issues?	
How is the building lighting controlled? • Always on • Manually controlled • Occupancy sensor • Computer control • Indicate where each is used	

Maintenance

How often are filters changed? What types/efficiencies of filters are used?	
How often are air handler/furnace coils cleaned? Include both heating and cooling coils.	
How often are boilers and gas appliances checked for operating efficiency? This requires flue gas testing—can you provide printouts of the results?	
How often are boiler heat exchanger surfaces (fire side and water side) cleaned?	
How often are chiller condenser tubes cleaned?	
How often are building temperature controls calibrated?	
How often are VAV box controls checked to see if they are still 'alive'? (5-7 year life is typical)	
How often are control valves checked for positive seating to avoid internal leak-by?	
For water-cooled equipment (cooling towers, evaporative cooling, air washers, fluid coolers, etc.) do you have logs of water used for a 12-mo. period?	
Kitchen—How often are refrigeration evaporator (cold) coils and condenser (heat rejection) cleaned?	
Compressed Air—How often are compressed air pipes, hoses checked for leaks?	
Steam and Condensate—How often are steam traps checked for internal leak-by?	
Data Center—how often are water cooled shell/tube heat exchangers cleaned? How often are dry cooler coils cleaned?	
Other heat exchangers on a Cooling Loop (such as ice machines, refrigeration machines, dedicated cooling units)—how often are water cooled shell/tube heat exchangers cleaned?	
Cooling tower—is the basin kept clean? Is the water in the sump clean?	
Fluid Cooler—heat exchanger coil pack ever cleaned?	
Plate-Frame heat exchanger—ever cleaned?	
Evaporative cooling/Air washers—what is the procedure and frequency for cleaning and sanitizing? Is the sump automatically drained and filled with fresh water? How often?	

HVAC/Mechanical

Thermal Break Even Temperatures: • Below what OA temperature is heating used? • Above what OA temperature is cooling used?	
What mechanical equipment and areas are noisiest? Hottest?	
Are there any areas that are humidity sensitive? e.g. clean rooms, etc.	
Are there any areas that have year-round 24x7 cooling requirements? Do these spaces cause the entire main HVAC system to run to cool a small area?	
Is spot-cooling or spot-heating used?	
If there are any areas with routine heating cooling problems (cold complaints, hot complaints) note them so we can see these areas.	
Are there any areas with indoor air quality (IAQ) complaints? Stuffy areas?	
Demand Controlled Ventilation (DCV)	
Is CO_2 control used to reduce outside air during low occupancy? If yes, is it measured by individual zone or is it measured in the return plenum for 'averaging'?	
Are there any areas that have requirements for simultaneous heating and cooling such as a dehumidification cycle or process control?	
Do you have any dual duct, multizone, or constant volume terminal reheat HVAC systems? If yes, note the locations so we can see these.	
Are there heating processes (water heating, HVAC, or process) that use electric resistance for heating? • Water heating • VAV boxes • Duct Heaters • Process tank heaters • Ovens Please indicate these so we can see them.	
Are there any areas that are air conditioned but do not have air-side economizers (outside air dampers for free cooling)? If yes, indicate where.	
Are there any humidifiers? • Electric • Gas	

(Continued)

• Steam What are they used for? • Winter comfort • Static electricity • Process	
Are there any evaporative cooling units? • Swamp coolers • Air washers What areas do they serve?	
Do you use fume hoods? Do they run all the time or are they turned off when not in use? Are there sash doors to reduce air flow when not in use?	
Are any of the air handlers using 100% Outside Air? If yes, note where they are so we can see them	
Are there transfer ducts or transfer fans used as make-up for exhaust only areas, such as restrooms, kitchen, etc?	
For VAV air handlers, how is the fan capacity accomplished? • inlet vanes • VFD • other	
HVAC main supply air temperature? Does the supply air temperature vary by season or stay the same?	
Are there any dedicated heating or cooling equipment for certain areas that are separate from the primary heat/cool systems, such as computer rooms units and other decentralized equipment?	
Are there any heat recovery systems? If yes, please indicate where. • Pre heating/pre cooling outside air? • Pre heating boiler feed water • Pre heating domestic hot water • Etc.	
Are there any air handlers that have been retrofitted with VFDs where the original inlet vanes are still there?	
If there is a water-cooling loop, please list the equipment served by this loop, it's approximate capacity or horsepower, and the temperature of the loop. Is the loop cooled by mechanical cooling or evaporative cooling (cooling tower or fluid cooler)? Is the cooling loop an "open" system or "closed" system? If open system, additionally indicate how often the heat exchangers are cleaned.	
For fan powered VAV boxes, are the box fans "series" (box fan runs all the time) or "parallel" (box fan runs only in heating mode)?	

Appendix

Central Mechanical Systems

Do you have a 'two pipe' systems? This is where the distributed piping carries either heating water or cooling water but not both.	
Chillers: • Year built • Air-cooled or water cooled? • Tons capacity • kW/ton or data sheets that show amps at full load cooling. • Is capacity modulation method by inlet vanes or VFD? • Are any of these are redundant units or do all the units run together during summer. • Do the units ever short-cycle during low load?	
Do you have absorption chillers? If yes, do they operate with waste heat or new-energy heating? Steam or hot water or direct fired?	
Hot Water Boilers: • Year built • Forced draft or natural draft (draft hood)? • Fuel? • Burner: 　—Start-stop? 　—Hi/lo fire? 　—Modulating? • Are any of these are redundant units or do they all run together during winter? • Do the units ever short-cycle during low load? • When a boiler is off (not firing), is the system hot water still being pumped through it? • Stack dampers used?	
What is the chilled water delta-T to the building? (The difference between entering and leaving temperatures).	
What is the heating water delta-T to the building? (The difference between entering and leaving temperatures)?	
Are there any 'jockey' boilers or chillers for mild weather use, when loads are low?	
Cooling Towers: • Year built • Fan Horsepower • Typical outlet temperature in summer? • Modulation method (on-off? two-speed? VFD?) • Describe water treatment that controls scaling and corrosion. • Cycles of concentration?	

Plate Frame heat exchanger: • Year built • Describe at what conditions this is used for free cooling. • Has the unit ever been cleaned?	
Any pumps or fans known to be way oversized, causing excessive throttling with dampers or balance valves to throttle the excess capacity?	
Chilled water pumping: • Constant flow or variable flow with VFDs?	
Heating water pumping: • Constant flow or variable flow with VFDs? • Mixing valve used for building water supply?	

Domestic Cold Water and Domestic Water Heating

How is domestic water heated? Does domestic water heating depend upon central boiler operation?	
Do points of use have low flow fixtures? • Showers (1.5 or 2.5 gpm) • Lavatories (0.5 or 1.0 gpm) • Water closets (1.6 gpf) • Urinals (0.5 or 1.0 gpf or waterless)	
If there are showers, can you estimate how many showers are taken per week? This may be a steady average or may be seasonal.	
Are there stack dampers on gas-fired water heaters?	
What temperature hot water is supplied?	

Other

Any other major equipment pieces that affect energy use.	
Anything that runs all the time that could be turned off sometimes?	
Anything that is kept at a constant temperature or pressure all the time that could be turned off sometimes?	

Controls

Operational issues with controls: • Temperature swings? • Hunting (oscillating back and forth) control? • When last were the instruments calibrated?	

(Continued)

How sure are you that control valves are closing tightly when called to close?	
How old are the control valves? • Main heating valves • Main cooling valves • Terminal reheat	
How old are the control dampers? • Return/outside air/relief economizer dampers • Inlet vane dampers	
Outside Air Economizer: • What is the air-side economizer control sequence? • Are there temperatures below which the economizer is locked out? • Any nuisance tripping of freeze stats from using the air economizer?	
What is the water-side (flat plate) economizer control sequence? • Does it activate based on dry bulb temperature (regular temperature) or wet bulb temperature?	
Scheduled Operation of Equipment? Indicated weekday control times vs. weekend control times. • Indoor lighting • Outdoor lighting • Air handlers • Exhaust fans • Chillers • Boilers • Pumps	
At night when air handlers shut down, do the associated exhaust fans shut down too?	
What are the control system set points for: • Zone temperature settings for heating season and cooling season, and deadband between heating and cooling? • Unoccupied set back temperature values? • Air Handler fan static pressure set points? • Variable pumping static pressure set points? • Fan tracking sequences of operation? • Supply air discharge temperature set point? Is this automatically reset? If yes, describe how the reset is done. • Cooling tower leaving water temperature set point? Is this a fixed set point or does it vary? If it varies, describe how it is done. • Temperature below which the chiller is locked out? • Temperature above which the boiler is locked out?	

(*Continued*)

• Supply air reset schedule. Reset from outside air, return air, other, or not at all? • Hot water reset schedule. Reset from outside air, other, or not at all? • Chilled water reset schedule. Reset from outside air, other, or not at all? • Demand Controlled Ventilation CO2 set point? • Air-economizer enable temperature set point	
Do you have logs of chiller (or main cooling units) run hours during the year and the corresponding outside air temperature? Do you have logs of boiler (or main heating units) run hours during the year and the corresponding outside air temperature?	
Zone override capabilities • Lights—how long? • Temperature control—how long?	
Do air handler controls that sequence preheat-mixed air-mechanical cooling have staggered set points or deadbands to prevent • heating-cooling overlap? • mixed air-mechanical cooling overlap? • evaporative cooling-mechanical cooling overlap? There should be a clear and defined dead zone between each function to be sure there is no overlap at all. • Or are overlaps allowed. Please explain.	
For VAV boxes, what are the following settings for five typical boxes: • Max cooling CFM • Min cooling CFM • Min heating CFM (if different than min cooling CFM)	
What other optimization routines are being used? • Optimal start? • Optimal stop? • Summer morning cool-down using outside air? • Winter morning warm-up with outside air temporarily closed? • Condenser water reset? • Chilled water reset? • Heating water reset? • Fan static pressure reset on VAV air handlers? • Demand Controlled Ventilation (reducing outside air proportionally when there are less people)? • Optimal chiller/boiler sequencing from efficiency curves/mapping?	

(Continued)

Appendix

- Variable air flow pressure reset from greatest individual demand?
- Distribution pump pressure reset from greatest individual demand?
- Load limiting of primary electrical equipment to 90%?
- Approach temperature monitoring for predictive maintenance of primary equipment?
- Other (describe)

SPECIAL USES

Swimming Pools
Restaurant/Kitchen
Laundry
Steam and Condensate
Compressed Air
Light Manufacturing
Computer Room/Data Centers

Swimming Pools

Hours of operation for each pool?	
Square foot surface area, volume, water and surrounding air temperature of each pool?	
How is humidity controlled in the pool area?	
Does the pool air connect to any other parts of the building besides the pool?	
Do you find a lot of condensation on the walls or roof?	
Do you cover the pool?	
Filtration: • Does the system use sand filters and blow down, or cartridge filters? • If sand filters, what is the blow down rate (gpm), times per week, and duration per event?	
Pool Heating: • How is the pool heated? • If gas heaters, how old are they, input rating, and efficiency rating? • Any condensing heaters used?	

(Continued)

Is the make-up water sub metered?	
Does the pool leak?	
Does air move from the lockers toward the pool, or the other way?	
Heat recovery for pool area exhaust to preheat incoming make-up air?	

Restaurant/Kitchen

Serving Breakfast/Lunch/Dinner?	
Meals per Day?	
Cooking Operations at What Fraction of Max Capacity? I.e. underutilized cooking equipment running?	
Hours of Operation: • Restaurant Hours for Customers? • Extended Store Hours for Staff?	
How many hoods?	
Hood Control—Exhaust and Make-up Interlocked to Run Together?	
Exhaust over Dishwasher?	
For each hood: • Hood used for? • Hood Control—When On/When Off? • Are Hoods Run Only When Cooking? • Ceiling temperature just outside of hood discolored or hot?	
Hood Make-Up Air • Cooling Tempered with DX or evaporative cooling or neither? • Heating Tempered with electric or gas heat or neither? How warm?	
Is the kitchen comfortable or uncomfortable during cooking periods?	
Indicate Gas or Electric • Grille • Fryer • Ovens	
Cooking Equipment • Cooking equipment turned down during slow periods? • Cooking equipment turned off at night?	

(Continued)

Appendix 413

- Hot equipment adjacent to and warming surfaces of refrigerated equipment?
- Food Heater Lamps Used? Do the lamps stay on even without food under them?

Indicate operating temperatures:
- Walk-in cooler(s)
- Reach-in cooler(s)
- Walk-in freezer(s)
- Reach-in freezer(s)

Walk-in cooler/freezers:
- Latches hold door closed tightly?
- Door gaskets in good condition?
- Doorway plastic curtains used to keep cold air inside?
- Door-bottom sweep used?
- Walk-in door for freezer located inside walk-in cooler?

Any equipment discharging warm air into the kitchen?
- Air cooled ice machines
- Air cooled reach-in cooler
- Air cooled walk-in coolers or freezers

Thermostats:
- Thermostat programmable?
- Settings set back at night?

Temperature settings:
- Kitchen-summer
- Kitchen-winter
- Is the kitchen area heated?
- Dining-summer
- Dining-winter
- Hood make-up air—summer
- Hood Make-up air—winter

Does the kitchen A/C unit have an economizer?

Kitchen Water Heating
- Water Heater—Gas or Elec?
- Stack damper?
- Water temperature?
- Does the dishwasher/pot washer use 180 degree hot water or chemicals for sanitizing rinse?

Laundry

Extraction:
- Centrifugal
- Press
- Percent extraction of washed clothes before drying

Waste water heat recovery to incoming water?	
Equipment turned off when not in use?	
Temperature of hot water as low as possible?	
Cooling system: • Air conditioning • Swamp cooler • Just exhaust	
Is heating provided in the laundry?	

Steam and Condensate

Year built	
Are any of the boilers redundant units or do they all run during periods of high demand?	
What is the steam pressure at the point of use?	
Steam Pressure: • Is steam being used at the lowest possible pressure? • Has an attempt been made to operate at reduced pressure? • Are there large uses that must be pressure-reduced from the main system supply that could otherwise be generated at a lower pressure? • Burner: —Start-stop? —Hi/Lo Fire? —Modulating?	
What is the steam used for? Comfort heating? Process heating/cooking? Humidification?	
What percent of the condensate is returned to the system? (e.g. condensate piping leaks?)	
Any heat recovery economizers for preheating boiler air or water?	
Is the boiler system turned off during unoccupied times?	
Stack dampers used?	

Compressed Air

(Note: this is for shop air, not pneumatic control air supply)

CFM and motor hp of each unit	

Appendix

Air Pressure: • What is the compressed air pressure at the point of use? • Are there large uses that must be pressure-reduced from the main system supply that could otherwise be generated at a lower pressure? • Is the system delivering air at the lowest possible pressure? • Has an attempt been made to operate at reduced pressure?	
What tasks that use air pressure could be served from low pressure 25psi blowers? • Blow off? • Other low pressure use?	
Is the outside air taken from the building or from a separate duct?	
Any heat recovery from the compressed air system?	
Is CA turned off when not in use? Is the compressed air system turned off during unoccupied times?	

Light Manufacturing

Do utility costs track production levels? i.e. at half production capacity, do utility costs lower to around half and follow the production rates up and down? If not, why?	
Are utility costs attributed to the manufactured product? i.e. the energy cost embedded in the part.	
Scheduling: • Do production schedules allow spreading out of the energy-intensive activities, to avoid heavy low load factors and high demand charges? • Are low load factors and high demand charges the result of Just-In-Time production?	
Are oven door seals tight fitting? Look for discoloration around seals and high temperatures.	
Stand By Loses: • Is heating equipment turned off when not being used, or left running in standby mode? • Other equipment turned off when not in use?	
Cooling in the manufacturing area: • Air conditioned • Swamp coolers • Spot cooling • Just exhaust	

Heating in the manufacturing area: • Are space heaters used and is this heat being lost to high bay areas instead of at the floor where workers are? • Spot heating	
Are heated tanks covered or open?	
Process exhaust fans: • Describe what each is used for. • Are there any that run continuously for fume hoods, paint booths, dust collection, etc. • Are these turned off when process is not active? • How are the individual exhaust drops controlled when the point of use is not active? Examples are a branch shut off damper, sash door, on-off switch, etc.	

Computer Room/Data Centers

Square footage of data center?	
Overhead lights on continuously or on an occupancy sensor?	
Temperature and humidity settings: • What are the temp/humidity settings? • Could relative humidity be lowered to 30%? • Could space temperature be raised to 72 degrees?	
Humidifiers: • Any humidification equipment other than what is in the CRAC units? • What separates the humidified areas and non-humidified areas? • Is there a tight vapor barrier to outdoors, adjacent floors and adjacent un-humidified spaces?	
Could chilled water set point be raised to 50 degrees?	
Could electric heat be removed?	
Unintentional dehumidification • Is condensate visible in the drains or on the cooling coils? • Are the coil faces wet to touch?	
Are heating and cooling mode lights ever on in opposite modes for adjacent equipment?	
Are humidification and dehumidification mode lights ever on in opposite modes for adjacent equipment?	

Appendix 417

Computer Equipment • Is the total computer equipment load known? • An equipment list may be useful, but sum of data plate values is more than actual load. Operator logs for a main UPS may provide this information if it serves all the computer equipment.	
Is the overhead lighting powered from the UPS?	
Type of CRAC units: • A—Air-cooled splits with compressor indoors and condenser outdoors. • B—Glycol fluid heat exchanger, compressor indoors and dry cooler outdoors. • C—Glycol fluid heat exchanger, compressor indoors and fluid cooler (or heat exchanger + cooling tower) outdoors. • D—Chilled water: No individual compressors, all cooling from air cooled chiller outdoors. • E—Chilled water: No individual compressors, all cooling from water-cooled chiller outdoors. • F—Other (describe)	
For chilled water CRAC units, what is the chilled water supply temperature?	
For down-flow CRAC units, is the floor insulated?	
Is there any connection between this space and the building HVAC system? If yes, please describe.	
For each computer room air conditioner or other cooling system dedicated to cooling a computer room. For identical units, fill out one and note quantity. • Unit ID • Make, model number, and year built • Nominal Capacity (tons) • Type of cooling equipment (A-F from above) • Enhancement: water-cooled economizer feature with an extra coil in the return air? • kW of electric heat? • kW of humidifier? • Type of humidifier built-into the unit (electric resistance canisters, infrared pan, ultrasonic, etc.) • Hours of operation—continuous? • Does this unit run at max capacity in summer? • Does this unit run at max capacity in winter? • Are there times this unit doesn't keep up? • Are there backup units, or do all the units run all the time? • Is the load consistent or does it vary throughout the day?	

FACILITY GUIDE SPECIFICATION:
SUGGESTIONS TO BUILD IN ENERGY EFFICIENCY

Source: "Strategic Facility Guidelines for Improved Energy Efficiency in New Buildings," Strategic Planning for Energy and the Environment (SPEE) Journal, Vol. 26, No. 3, 2007. Includes March 31, 2007 errata.

A guide specification is a hand-out document given to a design team at the beginning of a project to provide general instructions and owner preferences. The owner handing out the guidelines has more of an effect on the end result than the same suggestions made from a lone team member, hence the term top-down. These instructions are then integrated into the other governing documents and codes that eventually form the design. Traditional guide specifications, used by national accounts, campuses, and large facilities, spell out preferred manufacturers, acceptable types of piping, valves, light fixtures, pavement, etc. The concept of energy efficient guidelines in the owner's guide specification is a natural and overdue extension of an existing document. Even if the owner does not have a guide specification to add this to, the listed items in this document can be used in standalone fashion to serve the same purpose and provide the same benefit to the owner.

General:
- Design document submittals must include detailed narrative descriptions of system functionality, features, limitations, design assumptions and parameters, for use by the owner. The narratives will be detailed enough to provide benefit to subsequent design teams, and will be written to be informative and useful to building operations personnel. The narrative will be provided as a deliverable with the schematic design, and will be updated with each subsequent design delivery including DD and CD phases. In its final form, this document shall be placed on the first sheet of the drawing set behind the title page, so that the information is retrievable years later when all that is available to facility operations are the drawings. Design assumptions include number of people, indoor and outdoor HVAC design conditions, foot-candles of illumination, hours of operation, provisions for future expansion (if any), roof snow load, rainfall rates, etc. that all define the capabilities of the building.
- All equipment schedules, including HVAC, plumbing, lighting, glazing, and insulation shall be put onto the drawings and shall not reside in the specification books, so that the information is retriev-

able years later when all that is available to facility operations are the drawings.
- Design thermal insulation values and glazing properties that affect energy use (U-Value, shading coefficient, etc.) shall be clearly noted on the drawings.
- Project commissioning that includes identifying measurable energy savings goals and monitoring the design and construction activities with these as Project Intent items, with early detection and notification of any project changes that impact energy use or demand.
- Project final payment contingent upon:
  — Receipt of accepted accurate as-built drawings, with accuracy verified by owner and signed by the contractor.
  — Receipt of accurate and complete O/M manuals, with certified factory performance data, repair parts data, and vendor contact information for all energy consuming equipment, including all HVAC and lighting equipment and controls.
  — Receipt of test and balance report that demonstrates design intent is met for air, water, and ventilation quantities, showing design quantities, final adjusted quantities, and percent variance. This would include all VAV box minimum settings shown, including both heating and cooling balanced air quantities. This would also include any equipment performance testing that was specified for the project.
  — Verification by the owner that the test and balance settings include permanent markings so these settings can be preserved over time.
  — Receipt of on-site factory-authorized start-up testing for primary HVAC equipment including chillers and boilers, with efficiency and heat/cool performance figures and heat exchanger approach temperatures to serve as baseline. The submitted reports would include as a minimum heating/cooling output, gas/electric energy input, heat exchanger approach temperatures, water and air flows.
  — Receipt of control shop drawings with detailed descriptions of operation.
  — Acceptance testing of the automatic control system using the approved sequence of operation, and verification that the sequences are fully descriptive and accurate. Acceptance testing also includes review of the control system man-machine interface provisions to become familiar with each adjustable point in the sys-

tem. Acceptance is by the owner who will witness each sequence as part of the turnover training requirements.
- Building design must prevent negative pressure condition, unless safety considerations require it.
- Electric resistance space heating, air heating, water heating not allowed, unless there is no means to get natural gas to the site.
- Portable space heaters not allowed, unless required for an approved emergency measure.

Energy Use, Overall Performance:
- Using ASHRAE 90.1 or local Energy Code as a baseline, demonstrate through computer modeling that the building energy use will be at least 30% less than this value.

Irrigation Water Use, Overall Performance:
- Using standard Kentucky Bluegrass sod and average regional rainfall rates as a baseline, demonstrate that irrigation use for the property will be 50% or less of this value.

Test and Balance:
- Balance using "proportional balancing," a technique that strives to reduce throttling losses, which permanently energy transport penalties (pump and fan power).
- Any motor over 5 hp found to be throttled with a resistance element (valve or damper) more than 25% must be altered by sheave change or impeller trim to eliminate lifelong energy waste from excessive throttling losses.
- All 3-phase motor loads, including HVAC equipment, must include voltage balance verification as part of the TAB work. Voltage imbalance of more than (1) percent indicate unbalanced electrical service in the building and unacceptable efficiency losses.
- Vertical return air shafts serving multiple floors require a balancing damper at each branch outlet to proportion the return air by floor.
- Air flow performance testing for all ARI certified HVAC factory packaged unitary equipment greater than 5-tons capacity. Heating and cooling performance and efficiency verification is assumed via the ARI certification process.
- Heating efficiency, cooling efficiency, and air flow performance testing for all HVAC *split system* equipment greater than 5-tons capacity or 200,000 Btuh input heating capacity.

- Water flow performance testing for all ARI certified factory packaged water chillers. Cooling performance and efficiency verification is assumed via the ARI certification process.
- Water flow and combustion efficiency testing for all boiler equipment.
- Combustion efficiency testing for all boiler equipment unless factory startup is provided on site.
- Cooling tower thermal performance verification is assumed via the CTI certification process.

Electrical Service:
- Provide separate utility metering for electric, gas, and water for the building, separate from other buildings.
- Electrical transformer conversion efficiency not less than 95% efficient at all loads from 25% to 100% capacity. Dry-type transformers NEMA TP-1 compliant.
- Locate transformers in perimeter areas that do not require air conditioning for cooling.
- Power factor correction on large motor loads, for overall building PF of 90% or better. Large mechanical equipment can be provided with the correction equipment. If motor loads are segregated, this can be done at the switchgear.
- Arrange switchgear and distribution to allow metering of the following electrical loads (requires segregating loads):
  — Lighting.
  — Motors and Mechanical.
  — Plug Loads and Other.

Envelope:
- Orient buildings long dimensions E-W where possible to reduce E-W exposure and associated solar load.
- Provide building entrance vestibule large enough to close one door before the next one opens (air lock).
- Where thermal breaks are used, the thermal break material must have thermal conductivity properties an order of magnitude better than the higher conductivity material it touches, and must be at least 1/2 inch thick.
- *Minimum wall insulation <u>25% beyond ASHRAE 90.1 values, but not less than R-19.</u> Insulation is generally not expensive during new con-

struction. Incorporate exterior insulation system (outboard of the studs) for at least one half of the total R-value, to avoid thermal short circuits of standard metal stud walls, which de-rate simple batt insulation system by approximately 50%, e.g. a standard stud wall with R-19 batts between the studs yields an overall R-9.5.

- *Minimum Roof insulation R-value <u>25% beyond ASHRAE 90.1 values, but not less than R-30</u>. Insulation is generally not expensive during new construction. Select insulation that will retain its thermal properties if wet, e.g. closed cell material.
- Glazing meeting the following requirements:
  — Thermal breaks required.
  — *U-factor of 0.35 or less <u>where HVAC heating is provided</u>.
  — *Low-E coatings on East and West facing glass <u>where HVAC cooling is provided</u>.
  — *Max shading coefficient of 0.2 <u>where HVAC cooling is provided</u>. Note: any combination of tinting, coating, awnings or other exterior shading can be used to achieve this. This is to say that no more than 20% of the heat energy from the sunlit glazing is to get into the building.
  — Glazing not more than 25% of gross wall area.
- Skylight/Clerestory elements must meet the following requirements:
  — Thermal breaks required.
  — Triple pane (layer) construction with sealed air space(s)
  — Overall U-value of 0.25 or less.
  — *Skylight shading coefficient must be 0.2 or less <u>where HVAC cooling is provided</u>.
  — *Low-E coating <u>where HVAC cooling is provided</u>.
- Skylight/Clerestory area not to exceed 5% of roof area.
- Return plenums and shafts designed with an air barrier for leakage not exceeding 0.25 cfm/square foot of building envelope surface area @ 50 Pa (EBBA Criteria). Shaft construction requires field testing and verification.
- Building envelope devoid of thermal short circuits. Provide thermal break at all structural members between outside and inside surfaces.
- Building leakage testing required (new buildings), with no more than 0.25 cfm/square foot of building envelope surface area @ 50 Pa (EBBA Criteria).

Appendix

- Utilize lower ceilings to reduce necessary light input power for equivalent light levels at the work surface.
- Utilize reflective (light) color interior colors for ceilings, walls, furniture, and floors, to allow reduced lighting power for comparable illumination. It can take up to 40% more light to illuminate a dark room than a light room with a direct lighting system.
- Good reflectance parameters to use when picking interior surfaces and colors follow. If these values are used and the lighting designer is informed of it, the integrated design process will allow reduced lighting power to achieve the desired light levels.
  — Min 80% reflective   Ceiling
  — Min 50% reflective   Walls
  — Min 25% reflective   Floor and furniture
- Provide operable blinds for vision glass.

Lighting:
- Follow ASHRAE 90.1 or local Energy Code requirements for Lighting Power Budget Guidelines, and verify that designs are lower than these limits, while meeting current applicable IES lighting illumination requirements.
- Utilize task lighting and less on overhead lighting for desk work.
- Provide separate circuits for perimeter lights within 10 feet of the wall, to allow manual or automatic light harvesting.
- Use 1-2-3 switching for large open interior area spaces.
- Use ballast that will tolerate removing at least one bulb with no detriment.
- Where occupancy sensors are used, provide "switching ballast" that will tolerate large numbers of on-off cycles without bulb or ballast life span detriment.
- Use electronic ballast instead of magnetic ballast.
- Use ballast factor in the lighting design to improve lighting system efficiency. Because the ballast mostly determines how many watts are used, ballast choice is critical to achieving best energy efficiency.
- Coordinating light output with "ballast factor" is an excellent tool for providing optimum light levels and energy use.
- Use high power factor ballast, with minimum PF of 95% at all loads.
- Occupancy sensor in conference rooms, warehouses, and multi-function rooms. Also in locker rooms and restrooms, but with some continuous manual switched lighting in these areas.

- Photo-cell controlled lights in the vicinity of skylights.
- Do not use U-tube fluorescent lights, due to high replacement bulb costs.
- Do not use incandescent lights.
- Outdoor lighting on photocell or time switch.

Motors and Drives:
- All motors meet or exceed EPACT-1992 efficiency standards.
- VFD on all HVAC motors larger than 10 hp that have variable load.
- Motor nameplate HP not more than 20% higher than actual brake horsepower served (i.e. do not grossly oversize motors).

HVAC:
- Provide HVAC calculations and demonstrate equipment is not oversized. Equipment selection should not be more than 10% greater capacity than calculated values indicate.
- HVAC calculations will include both maximum and minimum heat/cool loads and equipment shall be designed to accommodate these load swings, maintaining heat/cool efficiency equal to or better than full load efficiency at reduced loads down to 25% of maximum load, e.g. equipment capacity will track load swings and energy efficiency will be maintained at all loads.
- Provide necessary outside air, but no more than this. Excess ventilation represents a large and controllable energy use. Reduce exhaust to minimum levels and utilize variable exhaust when possible instead of continuous exhaust. Reduce 'pressurization' air commensurate with building leakage characteristics. If the building is tested to low leakage as indicated herein, there should be little need for this extra air, or the heat/cool energy it requires. Design controls to dynamically vary outside air with occupancy.
- VAV box primary heating CFM shall be not higher than the cooling minimum CFM. This is to say the VAV box primary damper will NOT open up during heating mode.
- Zoning:
  — Design HVAC zoning to require heating OR cooling, not both. This will improve comfort and also reduce the inherent need for simultaneous heating and cooling.
  — Do not zone any interior areas together with any exterior areas.
  — Do not zone more than 3 private offices together.

Appendix 425

- — Do not zone more than one exposure (N, S, E, and W) together.
- Design and control settings for ASHRAE Standard 55 comfort envelope, which indicates 90% occupant comfort. Appropriate temperatures will vary depending on humidity levels. For example, in Colorado Springs (dry) the following space temperatures are appropriate:
    - — 71 deg F heating
    - — 76 deg F cooling
    - — Facilities may institute a range 68-72F heating and 74-78F cooling, provided a 5 degree deadband is kept between the heating and cooling settings.
- Do not heat warehouses above 60 deg F.
- Do not cool data centers below 72 deg F
- Do not use electric resistance heat.
- Do not use perimeter fin-tube hydronic heating.
- *In cooler climates* where HVAC economizers are used, designs should normally favor air-economizers over water-economizers since the efficiency kW/ton is better for the air system. The water economizer 'free cooling' includes the pumping and cooling tower fan horsepower, as well as the air handler fan. If the air handler fan power is considered required regardless of cooling source, the air-side economizer is truly 'free' cooling.
- *In very dry climates, with outdoor air wet bulb temperatures consistently less than 52 degrees and dew point consistently less than 42 degrees,* evaporative cooling (direct, indirect, or direct-indirect) should be used in lieu of mechanical refrigeration cooling, as long as indoor humidity of 40%rH or less can be maintained. To the water consumption issue, it is this author's opinion that water is a renewable resource and does not disappear from the planet like fossil fuels do, and so this technology should be used without environmental resources concern.
- Packaged HVAC cooling equipment not less than SEER-13 or EER-12, as applicable
- *Air-side economizers for all rooftop equipment, regardless of size, *for climates with design wet bulb temperatures below 65 deg F.*
- Avoid duct liner and fiber-board ducts due to higher air friction and energy transport penalties.
- Insulate all outdoor ductwork to R-15 minimum.
- Use angled filters in lieu of flat filters, to reduce air friction loss.

- Reduce coil and filter velocities to a maximum of 400 fpm to lower permanent air system losses and fan power.
- Avoid series-fan-powered VAV boxes.
- For fan-powered VAV boxes, use ECM motors to achieve minimum 80% efficiency. Although the motors are not large, when there are many of them this efficiency benefit is significant.
- Heat recovery for any 100% outside air intake point that is greater than 5000 cfm when the air is heated or cooled.
- Air filter requirements:
  — Terminal units (fan coils, fan powered boxes, unit vents): 20% (1-inch pleated) Note: this may require an oversize fan on small terminal equipment, and not all manufacturers can accommodate.
  — Air handlers with 25% or less OA: 30%—MERV-7
  — Air handlers with 25-50% OA: 45%—MERV-9
  — Air handlers with more than 50% OA: 85%—MERV-13
  — Provide manometers across filter banks for all air handlers over 20 tons capacity. Equip manometers with means to mark the "new-clean" filter condition, and change-out points.
- *Air cooled condensing units over 25 tons, provide evaporative pre-cooling, *for climates with design wet bulb temperatures below 65 deg F.*
- Make-up meter for all hydronic systems to log system leaks and maintain glycol mix.
- Separate systems for 24-7 loads to prevent running the whole building to serve a small load.
- *Direct evaporative post cooling for all chilled water systems, *for climates with design wet bulb temperatures below 65 deg F.*
- Require duct leakage testing for all ducts 2 in. w.c. design pressure class or greater.
- For process exhaust and fume hoods, design for variable exhaust and make-up.
- Utilize general exhaust air as make-up for toilet exhaust and other exhaust where possible.
- Dedicated outside air system (DOAS) for large office facilities (over 50,000 SF) with VAV systems, allowing ZERO minimum settings for all VAV boxes. This will eliminate the VAV reheat penalty, and eliminate the internal zone over-cooling effect from VAV minimums which often requires running the boilers throughout the year for comfort control.
- Separate interior and exterior VAV zoning for open-plan rooms to

utilize zero-minimums in the interior spaces.
- Do not use grooved pipe fittings in hydronic heating or cooling piping systems to prevent operating central heating and cooling equipment year-round on account of these fittings.
- Verify that all manufacturer's recommended clearances are observed for air cooled equipment.
- Humidification:
  — Do not humidify any general occupancy buildings such as offices, warehouses, or service centers.
  — In data centers ONLY, humidification should not exceed 30-35% rH.
  — Where humidification is used, humidifiers should be ultrasonic, mister, or pad type and should not be electric resistance or infrared type.
  — Do not locate humidifiers upstream of cooling coils, to avoid simultaneous humidification—dehumidification.
  — Where humidification is used, provide for elevated apparatus dew point of cooling coils or other means to prevent simultaneous humidification—dehumidification.
- Dehumidification:
  — Do not dehumidify below 45% rH.
- Provide performance and efficiency testing of package heating and cooling equipment over 7000 CFM or 20 tons or 500,000 Btu input heating units with factory authorized equipment representatives. Test figures to include on-site gross heat/cool output, fuel and electrical input, and efficiency, compared to advertised values.

Energy Transport Systems—Energy Budget:
- For HVAC air systems, the maximum energy transport budget will be:
  — No less than 10 Btu cooling and heating delivered to the space per Btu of fan energy spent at design conditions.

  This will generally steer the design toward generous sizing of sheet metal ducts, air handler cabinetry, coils and filters, higher efficiency fans (0.7 or better),and higher system differential temperatures to reduce air flow rates, but will result in greatly reduced lifetime energy use since it lowers the bar of system pressure.

- Fan hp limitation from:
  Cooling fan hp max input = Cooling Btu gross output/(10 * 3413 * motor-eff)

- Air hp limitation from:
  Cooling fan hp max budget * fan-eff.

- TSP limitation from:
  TSP = (air-HP * fan-eff * 6360)/CFM

For example, a 100-ton HVAC air system using 80%e motor, 70%e fan, and 350 cfm per ton would be limited to 44 hp motor load and 3.9 inches w.c. total static pressure.

NOTE: for systems with both supply and return fans, the transport energy considers both combined as the "fan"

- For HVAC water systems, the maximum energy transport budget will be:
  — No less than <u>50</u> Btu cooling and heating delivered to the space per Btu of pump energy spent, at design conditions.

This will generally steer the design toward generous sizing of piping, strainers, coils, and heat exchangers, higher efficiency pumps (0.75 or better) and higher system differential temperatures to reduce water flow rates, but will result in reduced lifetime energy use since it lowers the bar of system pressure.

- <u>Pump hp limitation from:</u>
  Cooling pump hp max input = Cooling Btu gross output/(50 * 3413 * motor-eff)

- <u>Water hp budget from:</u>
  Pump max hp * pump-eff.

- <u>HEAD limitation from:</u>
  HEAD = (water-HP * pump-eff * 3960)/GPM

Hydronic Circulating Systems:
- Heating: minimum 40 degree dT design, to reduce circulating flow rates and pump HP.
- Cooling: minimum 16degree dT design, to reduce circulating flow rates and pump HP.

Boilers and Furnaces:
- No atmospheric burners.
- No standing pilots.
- Design hydronic system coils to return water to the boiler at or below 140 degree water with a minimum of 40 deg temperature drop. This will reduce circulating pump energy and improve boiler efficiencies.
- Minimum efficiency of 85% at all loads down to 25% load.
- For heating load turn-down greater than 4:1, provide modular boilers or a jockey boiler.
- For multiple boilers sharing multiple pumps, provide motorized valves to cause water flow to occur ONLY through the operating boiler.
- Provide stack dampers interlocked to burner fuel valve operation.

Chillers:
- *Water-cooled centrifugal efficiency 0.5 kW/ton or less with 70 deg F condenser water and 45 deg F chilled water *for climates with design wet bulb 65 deg F and lower. 0.58kW/ton or less with 85 deg F condenser water and 45 deg F chilled water in climates where design wet bulb temperatures are above 75 deg F.*
- *Water-cooled centrifugal units able to accept 55 degree condenser water at 3 gpm per ton, all loads. *Beneficial in dry climates with design wet bulb temperatures less than 65 deg F and typical wet bulb temperatures less than 50 deg F.*
- *Water-cooled positive displacement units 0.7 kW/ton or less with 70 deg F entering condenser and 45 deg F chilled water *for climates with design wet bulb 65 deg F and lower. 0.81kW/ton or less with 85 deg F condenser water and 45 deg F chilled water in climates where design wet bulb temperatures are above 75 deg F.*
- Do not provide chilled water temperatures less than 45 deg F. Select cooling coils to provide necessary cooling with 45 deg F chilled water or higher.

- Air cooled chiller efficiency 1.0 kW/ton or less with 95 deg F entering air.
- *Air cooled chillers over 25 tons, provide evaporative pre-cooling *where design wet bulb temperatures are less than 65 deg F*.

Cooling Towers:
- Selected for 7 degree approach at design wet bulb and 0.05kW/ton or less fan input power. This will steer the design toward a larger free-breathing cooling tower box with a small fan, minimizing parasitic losses from the cooling tower fan. Cooling tower fan kW/ton should not be more than 1/10th of the chiller it serves.
- *Set condenser water temperature set point to no higher than 70 deg F *for climates with design wet bulb 65 deg F and lower. For climates with higher wet bulb temperatures, design to 7 degrees above design wet bulb with reset controls to lower the setting whenever conditions permit.*
- Water treatment control for minimum 7 cycles of concentration to conserve water.
- Specify cooling tower thermal performance to be certified in accordance with CTI (Cooling Tower Institute) STD-201.

Air-cooled Equipment and Cooling Towers in Enclosures:
- Locate to prevent air short-circuiting and associated loss of thermal performance. Rule of thumb is the height of the vertical finned surface projected horizontally. The fan discharge must be at or above the top of the enclosure and there should be amply sized inlet air openings in the enclosure walls as low as possible.

Ground Source Heat Pumps:
- COP 4.0 or higher at 40 deg F entering water.
- EER 17 or higher at 80 deg F entering water.
- No electric resistance heating.

Controls:
- Design OUT all simultaneous heating and cooling through the use of proper zoning, interlocks, and deadbands. This includes all constant volume systems and terminal unit systems. VAV systems inherently have an overlap which should be minimized by water and air reset in heating season, prudent use of minimum VAV box settings, and consideration of systems that separate the outside air from the supply air.

Appendix 431

- Programmed start-stop for lighting and HVAC systems, with option for temporary user overrides. Use these controls to prevent un-necessary operating hours.
- Lock out air flows for conference rooms and intermittent occupancy rooms by interlocking VAV box to close with occupancy sensors.
- Lock out chiller operation below 50 degrees, except for data centers or humidity-sensitive areas that cannot use outside air for cooling.
- Lock out boiler operation above 60 degrees, unless space temperatures cannot be maintained within the specified ranges any other way.
- *All cooling by air-economizer below 55 degrees *for climates with design wet bulb 75 deg F and lower.*
- Night set back for heating. Suggested temperature for unoccupied time is 60 deg F.
- No night set-up for cooling—no cooling operation in unoccupied times for general occupancy buildings. If building temperature rise during unoccupied times can cause detriment, then limit off-hours cooling operation to 85 degrees indoor temperature.
- Reset boiler hot water temperature settings in mild weather.
- *Reset chilled water temperature settings in mild weather, *provided outdoor air dew point is below indoor dew point levels.* Refrigeration savings generally exceeds increases in pump power.
- Provide appropriate interlock for all exhaust fans to prevent infiltration of outside air from uncontrolled exhaust fans that operate in unoccupied times.
- All analog instruments—temperature, pressure, etc. other than on-off devices—must be calibrated initially (or verified for non-adjustable devices). Merely accepting out-of-the-box performance without verification is not acceptable.
- 2-year guarantee on calibration, with 18-month re-calibration of all analog inputs.
- Air handler control valves with a residual positive seating mechanism for positive closure. Use of travel stops alone for this is not acceptable.
- For terminal units and heating/cooling hydronic water flow rates less than 10 gpm, use characterized ball valves for control valves instead of globe valves or flapper valves, for their inherent improved long-term close off performance. This will reduce energy use from simultaneous heating and cooling.

- Valve and damper actuator close-off rating at least 150% of max system pressure at that point, but not less than 50 psid (water) and 4 inches w.c. (air).
- Dampers at system air intake and exhaust with leakage rating not more than 10 CFM per square foot at 4" water column gage when tested in accordance with AMCA Standard 300.
- Water coil control valve wide open pressure drop sizing not to exceed the full flow coil water-side pressure drop.
- Provide main electrical energy and demand metering, and main gas metering. Establish baseline and then trend and log "kBtu/SF," "kWh/SF-yr," and "kW demand" perpetually and generate alarm if energy use exceeds baseline.
- Implement demand-limiting or load leveling strategies to improve load factor and reduce demand charges. Stagger-start any large loads, e.g. morning warm-up or cool-down events. Use VFDs to limit fan, pump, chiller loads to 90% during peak hours, etc.
- Independent heating and cooling set points for space control.
- Space temperature user adjustment locked out or, if provided, limited to +/- 2 degrees.
- 5 degree deadband between space heating and cooling set points to prevent inadvertent overlap at zone heat/cool equipment, and from adjacent zones.
- 5 degree deadband between air handler heating and cooling (or economizer) set points, e.g. preheat coil cannot share a single, sequenced, set point with the economizer or cooling control.
- Provide separate lighting and HVAC time schedules.
- For chillers (condenser) and hot water boilers, use temperature sensors to log heat exchanger approach values, to prompt predictive maintenance for cleaning fouled heat exchange surfaces. New-equipment approach will be the baseline value, and approach temperature increases of 50% will prompt servicing.
- Interlock heating and cooling equipment in warehouses serving doorway areas to shut off when roll-up doors are open to reduce waste.
- Optimization routines:
  — Automatically adjust ventilation rates for actual people count.
  — Optimal Start to delay equipment operation as long as possible.
  — Demand limiting control point that will limit all VFD-driven air handler fans components to a maximum of 90% max output in

summer. This will cause system temperatures to drift up slightly during extreme weather, but will reduce electrical demand for this equipment (and the cooling equipment it serves) compared to full output operation, during times when utility demand is highest. Do not oversize equipment capacity to compensate for this requirement.
- Optimal static pressure setting based on VAV box demand, not a fixed set point. This is a polling routine.
- *_For areas with design wet bulb temperatures below 65 deg F only,_ optimal supply air reset that will reset the supply air temperature set point upward from 55 to 62 for VAV systems during heating season, to reduce reheat energy. This can either be from two methods.
  - Method 1. Basic Optimization. When the main air handler fan is below 40% of capacity and OA temperature is below 40 degrees.
  - Method 2. Fully Optimized. Polling VAV boxes (at least 80% of the boxes served are at minimum air flows)
- Do not reset SA temperature from return air. Do not reset SA temperature during cooling season.
- Reset condenser water temperature downward when outdoor conditions permit, using the lowest allowable condenser water the chiller can accept.

Plumbing:
- Max shower flow 1.5 gpm
- Max bathtub volume 35 gallons.
- Max urinal water flow 0.5 gpf, or waterless.
- Max lavatory water flow 0.5 gpf.
- Metering (self closing) or infrared lavatory faucets.
- Avoid single lever faucets since these encourage complacency for the use of hot water.
- All domestic hot water piping insulated.
- Heat trap in domestic hot water main outlet piping.
- If a circulating system is used, provide aquastat or timer to prevent continuous operation.
- Max domestic hot water temp for hand washing 125 degrees.
- Gas water heaters in lieu of electric where natural gas is available.
- Domestic water heater equipment separate from the building boiler and heating system, to prevent year-round operation of central heat-

ing equipment.
- Water fountains instead of chilled water coolers.
- Operate the building at reduced pressure (such as 50 psig) instead of 70 psig, to reduce overall usage. Verify that design maintains at least 10 psig over the required minimum pressure at all flush valves.

Management and Maintenance Activities to Sustain Efficiency:
- Management Support
  — Create buy-in from the building occupants. Distribute information to building occupants to raise awareness of energy consumption, especially communicating that the user's habits are an essential ingredient to overall success, and are useful and appreciated. This would be in the form of occasional friendly and encouraging reminders of how user participation is helping, fun facts, etc., along with estimated benefits from behavior changes. Provide measured results whenever available.
  — Enforce temperature setting limitations, including the explanation of why this is helpful and also why it is reasonable. Encourage seasonal dress habits to promote comfort and conservation together.
  — Prohibit space heaters.
  — For offices, utilize LCD monitors and the software-driven "monitor power-off" feature, since the monitor represents 2/3 of the whole PC station energy use.
  — Track monthly energy and water use and maintain annual graphing lines, comparing current and prior years. Establish new benchmark curves after major renovations, alterations, or energy conservation projects. Compare annual use to benchmark and verify building energy and water usage per SF is not increasing. Report results to the building occupants as an annual energy use report for their feedback.
  — Escrow (save) approximately 5% of the replacement cost per year for the energy consuming equipment in the facility that has a normal life cycle, such as HVAC systems, lighting systems, control systems. This will allow 20-year replacement work without 'surprises' to sustain efficient building operations.
  — For leased office space, show the tenants their utility costs to increase awareness and encourage conservation by the users. The typical industry arrangement is to build in utilities into the

lease price, and so the tenants do not see a separate utility bill. Although the customers are paying for the utilities, having those costs clearly shown will reduce the complacency in utility use.

- Chillers:
  — Owner provide annual equipment "tune up" including cooling efficiency testing and heat exchanger approach measurements.
  — Owner adjust temperature settings or clean heat exchangers or adjust water flows whenever cooling efficiency tests are less than 90% of new-equipment values. For example, if new equipment benchmark is 0.5 kW/ton, then a measurement of 0.5/0.905=0.55 kW/ton would trigger corrective action.

- Boilers:
  — Owner provide annual equipment 'tune up" including combustion efficiency testing and heat exchanger approach measurements.
  — Owner adjust temperature settings, clean heat exchangers or adjust air-fuel mixture whenever combustion efficiency tests are less than 95% of new-equipment values. For example, if new equipment benchmark is 80%e, then a measurement of 0.8*0.95=0.76 would trigger corrective action.

- HVAC air coils:
  — Owner change filters at least quarterly, and verify there are no air path short circuits allowing air to bypass the filters.
  — Owner clean HVAC coils whenever there is any sign of visible accumulation or if air pressure drop is found to be excessive.

- HVAC air-cooled condensers:
  — Owner provide location free from debris, leaves, grass, etc. and adequate spacing for free 'breathing' and no re-circulation.
  — Owner clean heat exchange surfaces annually.

- Controls
  — Owner re-evaluate system occupancy times each year, to reduce un-necessary HVAC and lighting operating hours.
  — Owner re-evaluate control set points each year including space temperature settings, duct pressure settings, supply air temperature settings, reset schedules, heating and cooling equipment lock-out points.

— Owner re-calibrate control instruments each two years other than on-off devices.
— Owner cycle all motorized valves and dampers from open to closed annually, and verify tight closure.
— Owner cycle all VAV box dampers from open to closed annually and verify the control system is responsive, since these often have a short life and can fail without the user knowing it.

Footnote
1. This will no doubt be debated by some designers in semi-arid climates. However, experience has shown that at moisture levels much above these values it becomes increasingly hard to *guarantee* comfort all the time. Since the choice to use evaporative cooling is a large fork in the road for the customer, presenting conservative parameters is deliberate and are sure to work. If evaporative cooling is chosen in 'fringe' climates, a supplemental conventional cooling system integrated into the first stage of evaporative cooling is suggested.

Appendix

ASHRAE PSYCHROMETRIC CHARTS 1-5

Source: ASHRAE, © American Society of Heating, Refrigerating and Air-Conditioning Engineers, Inc., www.ashrae.org.

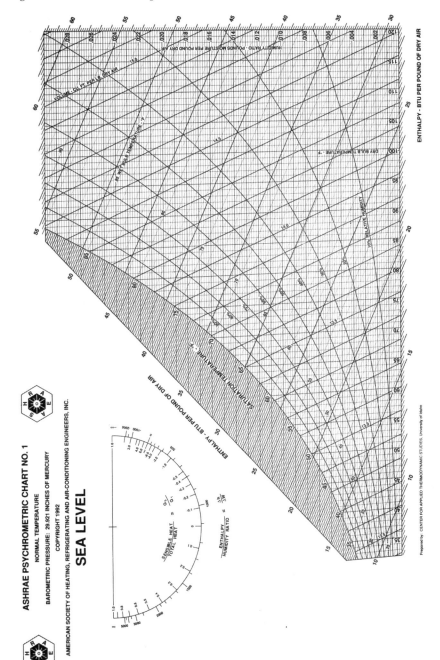

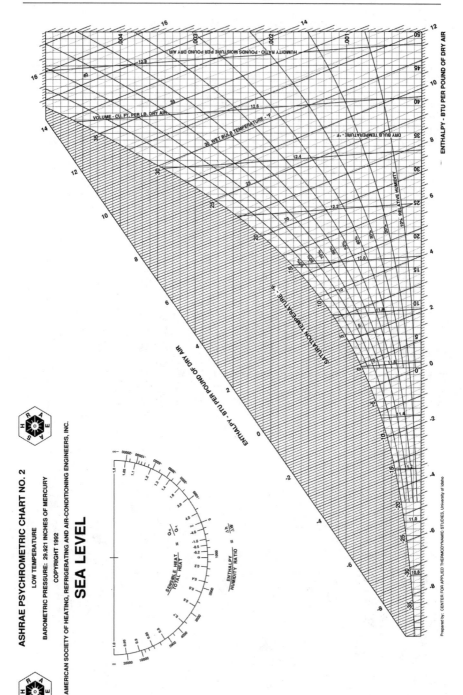

Appendix

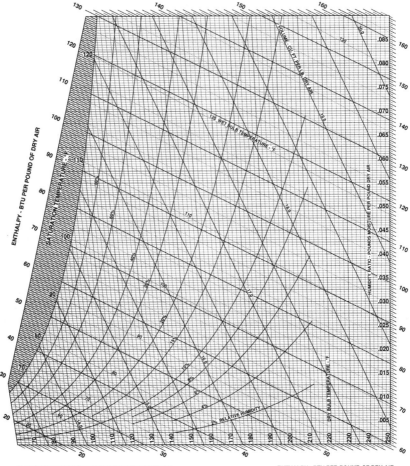

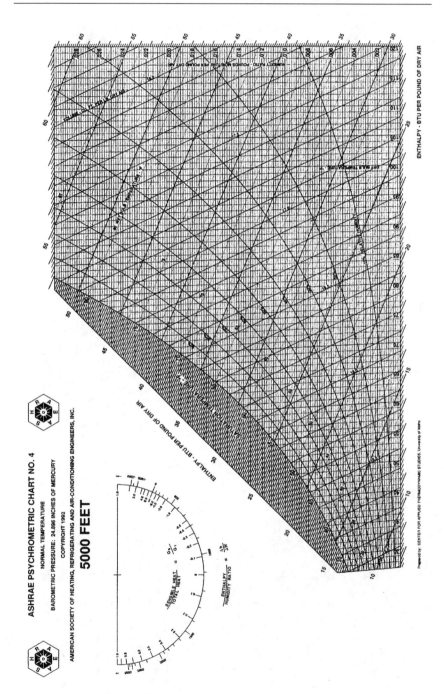

Appendix

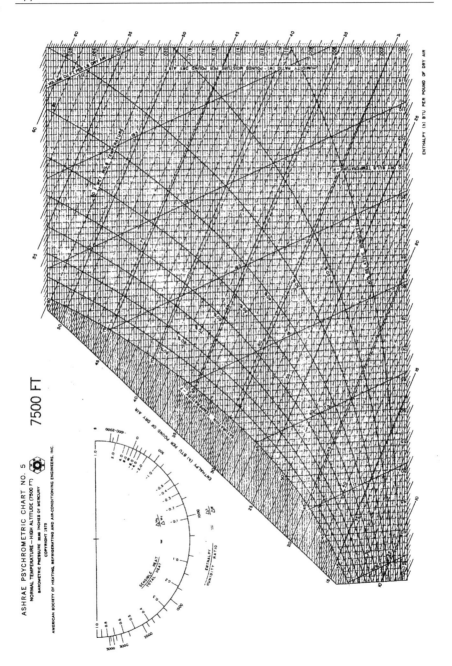

Index

A
absorption chiller 226
addressable lighting ballast 127
adiabatic cooling 27, 248
adiabatic humidification 82, 247
affinity laws 133, 161
    formulas 328
    modified for VFD savings 296
air-cooled equipment 430
air-side economizer 78, 184, 240
air-source heat pumps 250
air-to-air heat pumps 250
air and water circulating system
    resistance 235
air changes per hour 332
air compressor efficiency 289
air density ratios 334
air duct leaks 77
air economizer 167, 171
    savings 243
air horsepower 90, 208, 235
air re-entrainment 83
air recirculation 26
    air-to-air 26
    ground source (GSHP) 252
    heat pump 26
    water-to-air 26
air velocity on comfort 262
altitude correction 332
    factors 334
    factors at different temperatures 363
angled filters 80
anti-stratification fans 68
apartments 21
apparatus dew point 244

approach 66, 69, 135, 142, 143, 144, 176, 179, 186, 233, 430
    temperature 26
ASHRAE 90.1—items required for all compliant methods 379
auxiliaries 227
awnings 74

B
bag filters 80
ballast 297
    electronic 75
    factor 76, 298
    magnetic 75
baseline 213, 214
belt drive efficiencies 237
benchmark 3, 66, 233
beneficial heat 204, 225
bi-level switching 299
bin weather 119
    data 316
    data, cautions for using 317
    data for 5 cities (dry bulb) 362
BLC 305
    and cooling loads 306
blowdown vs. cycles of concentration 311
boiler 58, 201, 277, 429, 435
    blowdown 286
    feed water 97
    hp 279, 340
    isolation valves 281, 282
    isolation 282
    isolation valves 69
    lockout from outside air temperature 124

short cycling 280
skin loss 279
standby heat loss 279
brake horsepower 235
break-even temperature 156, 241
building-related illness 138
building leakage testing 422
building load coefficient 305
building use categories defined (CBECS) 371
bypass deck 81

C
calibrating and testing the computer model 210
carbon dioxide 321
cartridge filters 80
centrifugal chiller 226
CFL 76
CFM 332
check numbers 210
    for cooling and heating design loads 211
chilled water reset 124
    constant flow pumping 124
    variable flow pumping 124
chilled water system 225
    auxiliaries 227
    performance 228
chiller 58, 200, 225, 429, 435
    efficiencies 226
    lockout from outside air temperature 124
churches/worship 24
cleaning
    boiler fire tubes 140
    boiler water tubes 140
    chiller condenser tubes 140
    chiller evaporator tubes 140

condenser coils 139
    evaporator air coils 139, 140, 141
clerestories 75
climate zone map 370
CO_2 sensor 131
coefficient, duct fitting loss 91
cog belts 294, 295
coils, dirty 139
coincident wet bulb temperature 185
cold deck 80
cold duct 81, 133
colleges 28
color, exterior 73
combustion air 104
combustion efficiency 69, 278
    for some equipment 277
combustion temperatures 103
commissioning 382
compact fluorescent 76
compressed air 271, 398, 414
    reduced pressure 289, 414
cooler inlet air, heat recovery 290
cost 289
    inlet temperature 103
    leak 289
    pressure 103
    systems 70
computer data centers 26
computer equipment 26
computer modeling 153, 154, 210, 214
computer room 416
    air conditioning (data center) 244
    energy, proportion from cooling 245
condenser water reset 129, 176,

179, 180
condensing temperature 78
conduction heat flow 331
conflicting ECMs 345
constant flow pumping 124
constant volume 85
    systems 166
    terminal reheat 85
construction cracks 73
controls 408, 430, 435
control system cost/benefit 112
control valves 78
conventional cooling 257
conversion factors 327, 339
    energy 339
conveyors 69
cooling degree day 315
cooling tower 233, 430
    relative capacity factor 179
cool storage 99, 100
COP 252, 327
cost estimating—accuracy levels defined 354
cost vs. benefit 153
cracks, construction 73
CRAC unit 249
cubic feet per minute 332
cycles of concentration 311, 430

D

damper travel 79
data center 26, 247, 416
daylight harvesting 127
de-lamping 68, 77
dead band 115, 121, 157, 380, 432
dedicated outside air system (DOAS) 223, 263, 426
degree day, heating 315
    cooling 315
    base temperature 315
    cautions for using 315
dehumidification 158
delta T 78, 84, 85, 133, 330
demand charges 106, 122, 126
demand controlled ventilation (DCV) 131
desiccant 98
dew point temperature 158
differential temperature 85, 330
digital control system 111
direct evaporative cooling 255, 258
    post cooling 257, 261
dirty coils 139
discharge dampers 87, 89
district cooling 133
district heating 91, 133
documentation 215, 401
domestic water heating 311, 408
door seals 66, 69
dormitories 21
double use of process air and water 98
drive losses 235
dry cooler 26, 70, 234
dual duct system 222
dual duct terminal unit 133
duct and fitting pressure losses using "C" factor 337
duct and fitting pressure losses using equivalent diameters (L/D) 337
duct fitting 90
    and system pressure losses, using "C" factor 337
    loss coefficients 91, 359
duct friction 89
duct leaks 77

E

ECM motors 426
economic savings 241
economizer 20, 123, 171
    air-side 78
    cutoff point 240
    water-side 79
education, colleges and universities 28
    schools K-12 31
EER 327, 337, 338
    from nameplate compressor full load amp 337
effect of cooling tower energy on overall energy savings 180
efficiencies, belt drive 237
efficiencies, fan and pump 237
efficacy 298
electrical formulas 328, 329
electric resistance heat 126
electric resistance humidifiers 246
electronic ballast 75
electronic expansion valves 234
electrostatic discharge (ESD) 247
encapsulated ice 100
energy, end use distribution pie diagram 6
energy audit
    approach 395
    look-for items 397
    sample questionnaire 401
    types 348
energy consumption for heating outside air 267
energy conversion factors 339
energy transport 224, 330
    loss 172
    systems, energy budget 427
energy use graphs 9
energy use index 4
energy use intensity (EUI) 4, 5
    by function, climate, zone, and size 365
    by function and climate zone 365
    by function and size 364
    for some manufacturing operations 375
    measured at a data center 370
energy use intensity (EUI) - in production units, for some manufacturing operations 377
energy use profiles 9
enthalpy sensors 123
entrance losses 90
envelope 125, 305, 403, 421
    leakage 267
equating energy savings to profit increase 390
equipment spacing 83
eroding savings 125, 215
EUI 4, 5, 364, 365, 370, 377, 378
    calculating 4
    mixed 5
    production 5
evaporating temperature 78
evaporation 68, 93
    loss from water in heated tanks 361
evaporative cooling 67, 255, 256, 257
    direct 255, 258
    direct, post cooling 257, 261
    indirect-direct 256, 259
    indirect-direct with supplemental conventional cooling 257, 260
evaporative pad 247

Index 447

humidifiers 247
evaporative pre-cooling 26, 70, 82
excess air 69, 104, 287
exfiltration 267
exit losses 90
exterior color 73
exterior screening 309
exterior shades 309

F
facility guide specifications 418
fan-powered VAV box 426
fan/pump motor work equation 236
fan and pump efficiencies 237
fan and pump throttling 238
fan coil unit 222
fan horsepower 161, 208
feed water 103, 104
filter, angled 80
filter, bag 80
filter, cartridge 80
filter, flat 80
fin-tube heat exchanger 144
flat filters 80
flat plate 185
    heat exchanger 79, 134
floating balls 68
flow chart 63
fluid cooler 26, 67, 70, 233
    adiabatic 247
    humidifiers
    infrared
    ultrasonic
food sales 34
food service/restaurants 50
foot-candles 298
forced draft 277
formulas 327

fossil fuel emissions 324
fouling 66, 69, 142, 144
    in a boiler 141
friction
    duct 89
    factor 337
    losses 208
    pipe 89
full storage TES 99
furnaces 429

G
general exhaust 98
generators 288
glass, exterior shades 309
glass, interior shades 309
glass, low E coating 308
glass, shading 308
glass, silk screen shaded 309
    exterior screening 309
    solar film 309
glass, thermal break 307
glazing 422
    frame 307
    high performance 307
    properties 307
global point sharing 135
glycol vs. efficiency 263
    effect on chilled water pumping 264, 265
    effect on chiller efficiency 264
    effect on heating water efficiency 264
graphing 9
graphs, energy use 9
gravity flue 277
greenhouse gas 321
    emissions by gas 324
    relationship to energy use 321

grocery stores 34
ground source heat pump (GSHP) 223, 252, 430
guide specifications, facility 418

H
harmonics 103
head pressure 26, 80
health care, hospital 36
heat-conversion factors 327
heat-cool overlap 159
heating-cooling overlap 133
heating and cooling, simultaneous 133
heating degree days 305, 315
heating surface area 279
heat exchange, wash water 97
heat exchanger 143
   approach 143
   fin-tube 144
   flat plate 79
   plate-frame 144
   shell and tube 144
heat loss from uninsulated hot piping and surfaces 358
heat pump 223, 250
   air-to-air 250
   ground source (GSHP) 252
   water heaters 313
   water-to-air 250
   water-to-water 252
heat recovery 68, 95
   equipment 98
   refrigeration system 96
   viability test 95
heat transfer formulas 330
heat transfer surfaces 215
heat wheels 98
HID 68, 76

high bay 76
   lighting 68
high rise 59
horsepower, air 90
hospital 36
hotels 44
hot deck 80
hot duct 81, 133
hot water recirculation 314
hot water reset 204
   from zone demand 132
humidification 244, 427
   and dehumidification, simultaneous 244
   formula 334
humidifiers 70, 247
   adiabatic 247
   electric resistance 246
   evaporative 247
   fogging 247
   infrared 246
   ultrasonic 247
humidity-sensitive 242, 244
humid climates 157, 267
HVAC 380
   control 114, 127
   settings 115
   formulas 331
   retrofits for the three worst systems 385
hydraulic diameter 337

I
ice rinks 61
ice storage 102
idle mode 63
idling 400
idling equipment 18, 400
impeller trimming 86

Index 449

incandescent 76
indirect-direct evaporative cooling 256, 257, 259
indirect-direct with supplemental cooling 257
indoor air quality 137
   costs 137
indoor comfort settings 114
induction system 222
infiltration 73, 267
infrared humidifiers 246
inlet vanes 87, 89, 406
instantaneous water heaters 314
insulation 74
insulation formulas 335
integrated design 394
integrated part load value 228
interior shades 309
internal gains 242
   heat 239
internal loads 156, 186
internal rate of return (IRR) 355
IPLV 228
IPMV 212

J
just-in-time manufacturing 67

K
k-12 schools 31
kBtu 4
kBtu/SF-yr 3
k-value 336
kitchen 249, 412
   hood exhaust 249
   hood make-up 249
kW/ton 327

L
labor costs 66

latent heat of water 335
laundries 41
laundry 97, 413
leakage 267
   envelope 267
   testing, building 422
leaks, air duct 77
libraries/museums 42
lighting 75, 193, 194, 195, 297, 383, 403, 423
   ballast, addressable 127
   control 113, 127
   energy use, pct of total electric 301
   fluorescent 76
   HID 76
   hours by building type 301
   incandescent 76
   levels, typical recommended 302
   opportunities 302
   technology properties 300
   terms 298
light harvesting 75
load-following 159
load factor 15, 67, 108
   defined 106
   effect on energy cost 14
   limiting 122
   profile 63, 155, 159, 201, 294
   profile, HVAC 155
   profile, motor 201
   profile for humid climates 157
   profile for overlapping heating and cooling 159
lodging 44
loss coefficients, duct fitting 359
low-E coatings 308

M

M&V 212
   baseline 213
   options 213
magnetic ballast 75
maintenance 404, 434
   value 137
   energy benefits 129, 139
make-up air 68, 97, 126
make-up water 97
manufacturing 63, 415
measurement & verification 212
mercantile 53
minimum stack gas 279
mismatched compressor/evaporator coil 247
mixing box, zone 81
Mollier Diagram 253
most open valve 127
motels 44
motor 196, 269, 424
   efficiency 269
   efficiency at reduced load 271
   load profile 196
   losses 273
motor power factor at reduced load 275
multi-family buildings 21
multi-zone 80, 133
   hot deck/cold deck reset 132
   system 222
museums 42

N

normalize 214

O

occupancy data 214
occupancy patterns 132
occupancy sensor 120, 423
   control of HVAC 128
   energy savings 304
off-peak 105
office buildings 47
on-peak 105
operable window 125
operating budgets 215
operating costs 6
operating expense 6
operating expenses, percent from utility costs 375
optimization 111, 127, 432
optimum start 117
outlet damper 87
outside air 73
   energy consumption for heating 268
   reset of hot water converters 125
ovens 68
overhangs 74
overlapping heating and cooling 159, 268
owner-friendly design considerations 217

P

parasitic fan heat 26, 27
   loss 27, 98, 225
partial storage TES 99
part load refrigeration compressor power 229
part load benefits of variable flow pumping, chillers 230
part load boiler operation 280
part load chilled water system performance 228
people counter 131

percent occupancy 5
perimeter heating 307
perimeter lighting 127
phase change material storage 102
photo-cell 424
pipe (in) 358
pipe friction 89
plate-frame heat exchanger 144
plenum 27, 73, 77
plumbing 433
plumbing fixture water flow rates 312
polling 127, 129
pollution 321
pollution-conversion constants by region 322
pollution conversion to equivalent number of automobiles 325
pool 59, 93
    covers 93
    evaporation 94
pool air and water temperature 93
pool evaporation formula 94
power factor 102, 108, 328, 329
    correction 102, 274
    for some equipment 276
pre-cool 79
pre-heating 97
predictive maintenance 66, 135
pressure-temperature charts for refrigerants 349
pressure unit equivalents 340
preventive maintenance 137
process air and water, double use of 98
process analysis 63
process cooling 68
process flow chart 63
process flow diagram 64

production rate 16, 65
production scheduling 66
productivity increase 137
    equated to dollars 138
productivity value 138
profit 390
profit 6
programmable lighting ballast 127
project intent 215
properties of air, water, ice 334
proportional balancing 86
psychrometric charts 437
pumping, variable flow 181
pump horsepower 161, 184

Q

questionnaire, energy audit sample 401

R

R-value 335
    reduction from stud walls 306
ratchet clause 108
rate of return (ROI) 153
re-entrainment, air 83
recirculation, air 26
redundancy 210
reflectance factor 74
reflective values of common colors 299
reflectivity 73
reflectors 77
refrigerants, pressure-temperature charts 349
refrigerant hot gas 97
refrigerated warehouse 57
refrigeration compressor percent power at part load 229
refrigeration cycle 253

refrigeration efficiency degradation at part load 230
refrigeration system heat recovery 96
reheat 239
   energy penalty 166
   penalty 116
rejected heat 69, 97
relative cooling tower capacity factor 176, 179
relative efficiency of air conditioning systems 221
relief damper 78, 79
repair costs 137
reset
   chilled water, constant flow pumping 124
   chilled water, variable flow pumping
   condenser water 128
   hot water, from zone demand 132
   multi-zone hot deck/cold deck from zone demand 132
   schedule 125, 130, 204
   supply air, VAV from zone demand 128
   supply air static pressure 128
   supply water pressure, variable pumping 128
   VAV box minimum setting 128
   ventilation, from CO_2 131
   ventilation, by people count 131
restaurant 50, 412
retail/sales 53
return air plenum 73, 77, 399, 422
ROI 153

S

sales 53
scale 141
scaling 66
schools 31
SEER 327
sensible cooling 244
sequencing of multiple chillers/boilers 132
service access 216
service life of various system components 388
set back 117
   thermostat savings 120
set up/set back 117
   glass, thermal break
   exterior shades 308
   low-E coating
   interior shades
shading 307
   coefficient 308
   window 74
shell and tube heat exchanger 144
shift pay differentials 66
short cycling 400
   boiler 280
sick building syndrome 138
side-arm heater 280
silk screen shaded glass 309
simple payback 153, 355
simultaneous dehumidification/humidification 26, 27, 244
simultaneous heating and cooling 67, 133, 159, 268
single zone system 222
single zone VAV 223
skylight 75, 422
sleep mode 63
solar heat gain factor 308

solar load 306, 307
soot build-up 141
space heaters 434
specific heat 224
    of air and water 335
spot cooling 67, 82, 262
stack dampers 281, 282
stack gas temperature, minimum 279
stack temperature 103, 142
standby losses 19, 117, 125, 204, 313, 314, 400
    boiler 279
    water heaters 313
static pressure control setting 88
steam 414
    cost 277
    leak 284, 285
    pressure 103
    pressure, reducing 284
storage tanks for domestic hot water heaters 313
stranded cost 108
stratification losses 99
super T-8 76
supply air 129
    reset 166, 171
    reset of VAV systems 130
    static pressure reset 128
    temperature reset 129
supply water pressure reset 128
sustainability 224
swimming pools 93, 411
system resistance 89
    air and water circulating 234

T
TAB 86
tenant override 120
terminal reheat system 222
TES 99
    cool storage 100
    full storage 99, 100
    ice storage 100
    partial storage 99, 100
    phase change material (PCM) 100
    warm storage 100
testing adjusting and balancing (TAB) 86
Texas multi-zone 81
thermal break-even concept for buildings 239
thermal break-even point 130
thermal break-even temperature 154
thermal breaks 307
thermal conductivity 336
thermal resistivity 335
thermal short circuit 74
thermal storage 99
thermodynamic lift 26
throttling, fan and pump 238
ton-hours 100, 200, 332, 339
top 15 emerging technologies—2002 (DOE) 384
transport energy 161, 221
trimming, impeller 86

U
U-value 307, 336
ultrasonic humidifier 247
unit ventilator 222
universities 28
unoccupied mode 118
utility load tracking 135
utility rates 105

V

V-belt 294, 295
vapor barrier 27
vapor compression cycle 253
vapor pressure 93
variable air volume (VAV) 85
　systems 171
variable flow 231
　pumping 124, 181
variable pumping 181
variable speed drive 294
variable speed pumping 84
variable volume system 222
VAV 85
　fan savings de-rate factors 166
　reheat penalty 167, 263
VAV box 85
　box minimum air flows 171
　considerations 294
　cooling minimum air flow 116
　heating minimum air flow 116
　minimum air flow setting 399
　minimum settings 130
　minimum setting reset 128
　reheat penalty 171
　system fan savings reduction for maintaining downstream pressure 165
velocity head (Hv) 91
　formulas 235
viscosity, glycol in chilled water 266
ventilation 81, 267
　effectiveness by season 132
　reset from CO_2 131
vestibule 125
VFD 294, 424
　savings 296

viscosity, glycol 263
　in chilled water 266
voltage imbalance 270

W

warehouses 56
warm storage 99
waste water heat exchange 97
water-cooled 224
water-side economizer 79, 134, 184, 186
water-source heat pumps 252
water-to-water heat pumps 252
water flow rate when load and dT are known 182
water heaters 312
　storage tanks 313
water heating 382, 408
water horsepower 235
water source heat pump loop 224
water treatment 66
weather, bin 316
weather dependent 189
weather data, degree day 315
weather data by days and times 317
weather dependent 13
weather independent 14, 190, 192, 244
wet bulb depression 82
wet bulb temperature 116, 129, 176
window shading 74
work equation, fan/pump 236
worship 24

Z

zone mixing boxes 81
zone of greatest demand 114